黄风 编著

图解
工业机器人
技术英语

Diagrammatic English of
Industrial Robot
Technology

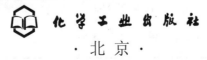

·北京·

Brief introduction

This book brings together the technical terms of industrial robots in various fields of practical applications. In the form of illustration, it is illustrated in both Chinese and English. There were more than 300 drawings. Technical terms include standard specifications, installation, setup, operation, programming, commands, parameters, advanced functions, tracking control, additional axis control, maintenance, inspection and other aspects of industrial robots. This book also introduces the applications of industrial robots in welding, transportation, polishing, loading and unloading of machine tool, cutting, spraying, as well as the latest technological progress of industrial robots. This is a reference book of great practical value for engineers, designers, maintenance technicians, teachers and students in universities who are engaged in the field of automation and robotics.

内 容 简 介

本书汇集了工业机器人在实际应用各领域中的技术术语，以图解的形式，用中英文两种文字进行了说明，计图 300 余幅。技术术语包括工业机器人的技术规格、安装、设置、操作、编程、指令、参数设置、高级功能、跟踪控制、附加轴控制、维护保养等方面。本书还介绍了工业机器人在焊接、搬运码垛、打磨抛光、机床上下料、切割、喷涂等各方面的应用以及工业机器人最新技术进展。

对从事自动化和机器人工作领域的工程师、设计人员和维保技师以及相关专业的高校师生来说，这是一本极具实用价值的参考工具书。

图书在版编目（CIP）数据

图解工业机器人技术英语/黄风编著．—北京：化学工业出版社，2020.10
ISBN 978-7-122-37263-5

Ⅰ．①图… Ⅱ．①黄… Ⅲ．①工业机器人-英语-图解 Ⅳ．① TP242.2-64

中国版本图书馆 CIP 数据核字（2020）第 106390 号

责任编辑：张兴辉　毛振威　　　　　装帧设计：王晓宇
责任校对：王佳伟

出版发行：化学工业出版社（北京市东城区青年湖南街 13 号　邮政编码 100011）
印　　刷：北京京华铭诚工贸有限公司
装　　订：三河市振勇印装有限公司
787mm×1092mm　1/16　印张 15$\frac{3}{4}$　字数 387 千字　2021 年 1 月北京第 1 版第 1 次印刷

购书咨询：010-64518888　　售后服务：010-64518899
网　　址：http://www.cip.com.cn
凡购买本书，如有缺损质量问题，本社销售中心负责调换。

定　　价：79.00 元　　　　　　　　　　　　　　　　　　　　　　版权所有　违者必究

前 言　Preface

20世纪60年代,在山城桂林的一个"小人书"摊前,一个小孩坐在小凳上看一本科幻的小人书,书中讲述了一个机器人冒充足球队员踢球的故事,这个冒名顶替的"足球队员"又能跑,又能抢,关键是射门准确,只要球队处于劣势,把他换上场就无往而不胜。这个故事太吸引人了,小孩恨不得自己就是那个机器人去征战球场。这个小孩就是当年的我。

60年过去了,有些科幻成了现实,有些现实超越了科幻。

现在工业机器人已经是机器人领域中的重要分支。近年来,工业机器人在制造领域的应用如火如荼,已经成为智能制造的核心技术。工业机器人行业是国家和地方政府大力扶持、重点倾斜的高新技术行业,工业机器人销量在全球所有主要市场均出现增长,2019年中国使用的工业机器人数量位居全球之首,将来会有更大的发展。

由于机器人技术是正在兴起的一门高新技术,使用者在学习和使用工业机器人的过程中,会遇到很多专业的技术术语。在现阶段使用的许多品牌的工业机器人的技术手册、编程手册、安装维护手册和故障排除手册中,很多是英文手册,这给学习和使用者带来许多不便,阅读英文手册时用户常常是一头雾水。本书作者根据多年使用及教学工业机器人的经验,将英文的工业机器人专业技术术语译成中文。这项工作既需要比较专业的理论知识,也需要丰富的现场实践经验。"信达雅"是个古老的话题，作者力图使这本书在"信达雅"方面有新颖完美的体现。

本书从实用的角度出发，以 300 余幅图片，使用中英文两种文字对工业机器人在技术规格、安装、设置、操作、编程、指令、参数设置、高级功能、跟踪控制、附加轴控制、维护保养等多方面的技术术语做了准确和详尽的解释。本书共有 14 章，第 1 章至第 11 章，是工业机器人的技术术语。第 12 章至第 14 章介绍了工业机器人在焊接、搬运码垛、打磨抛光、机床上下料、切割、喷涂等各方面的应用以及工业机器人最新技术进展。本书的读者对象是从事工业自动化和机器人工作领域的工程师、设计人员、操作人员、维保技师、设备管理工程师和高校自动化专业的师生。这是一本极具实用价值的参考工具书，把这本书摆在书案上或工具柜中是一个不错的选择。

作者学识有限，书中不免有疏漏之处，希望读者批评指教。

作者邮箱：hhhfff57710@163.com。

编著者

目 录 CONTENTS

Chapter 1　Standard Specifications of the Robot　机器人的技术规格 ……… 001

1.1　Robot arm　机器人本体 …………………………………………………… 001
1.2　Rated load and operating range　额定负载与动作范围 ………………… 002
　1.2.1　Relationship between mass capacity and movement area for 4-axis robot
　　　　4 轴机器人负载质量与动作范围关系 ………………………………… 002
　1.2.2　Relationship between mass capacity and movement area for 6-axis robot
　　　　6 轴机器人负载质量与动作范围关系 ………………………………… 003
1.3　Relationship between mass capacity and speed
　　负载质量与速度的关系 ……………………………………………………… 004
1.4　Relationship between height of shaft（J3 axis）and acceleration/
　　deceleration speed
　　J3 轴行程高度与加减速度的关系 …………………………………………… 005
1.5　Outside dimensions　外形尺寸 …………………………………………… 006
1.6　Tooling　工具（外围附件）………………………………………………… 007
1.7　Controller　控制器 ………………………………………………………… 010
1.8　Options　选件 ……………………………………………………………… 018

Chapter 2　Installing the Robot　机器人的安装 ……………………… 022

2.1　Unpacking to installation　开箱及安装 …………………………………… 022
2.2　Connecting the power cable and grounding cable
　　连接电源电缆和接地电缆 …………………………………………………… 023
2.3　Attachments installation procedures　附件安装 ………………………… 026

Chapter 3　Setting the Robot　机器人的设置 ………………………… 030

3.1　Setting the origin　原点设置 ……………………………………………… 030
　3.1.1　Origin data input method　数据输入方式 ………………………… 030
　3.1.2　Preparing the T/B　操作使用 T/B 示教单元 ……………………… 031
3.2　Resetting the origin　原点重新设置 ……………………………………… 032

3.2.1　Mechanical stopper method　机械挡块方式 …………………………………… 034
　　3.2.2　Jig method　校正棒方式 ……………………………………………………………… 037
　　3.2.3　ABS origin method　ABS 设置原点方式 ………………………………………… 039
　3.3　Changing the operating range　调整运行范围 …………………………………………042
　　3.3.1　The structure of mechanical stopper　机械挡块的构成 ……………………… 042
　　3.3.2　The installation of stopper block　机械挡块的安装 …………………………… 042
　　3.3.3　Installation of J1 axis operating range change option　J1 轴行程挡块的安装 …… 043

Chapter 4　How to Operate Robot　机器人的操作 …………… 045

　4.1　Operation panel（O/P）functions　操作面板的使用 ……………………………………045
　4.2　Installation of teaching pendant　示教单元的安装 ………………………………………047
　4.3　Explanation of operation methods　操作方式说明 ………………………………………048
　　4.3.1　JOINT jog　关节型点动 ……………………………………………………………… 049
　　4.3.2　XYZ jog　直交型点动 ………………………………………………………………… 053
　　4.3.3　TOOL jog　以工具坐标系为基准点动 ……………………………………………… 057
　　4.3.4　3-axis XYZ jog　三轴型点动 ………………………………………………………… 060
　　4.3.5　CYLINDER jog　圆柱型点动 ………………………………………………………… 063
　　4.3.6　WORK jog　在工件坐标系中的点动 ……………………………………………… 067
　4.4　Aligning the hand　机械手整列操作 ………………………………………………………073

Chapter 5　Programming　机器人编程 …………………… 075

　5.1　Creation procedures　编制程序的流程 ……………………………………………………075
　5.2　Creating the program　新建程序 ……………………………………………………………076

Chapter 6　Commands of the Robot　机器人编程指令 ………… 080

　6.1　Coordinate system description of the robot　机器人使用的坐标系种类 …080
　6.2　Mechanical interface coordinate system　机械接口坐标系 …………………………083
　6.3　Tool coordinate system　工具坐标系 ………………………………………………………084
　　6.3.1　Definition　定义 ………………………………………………………………………… 084
　　6.3.2　Effects of using tool coordinate system　使用工具坐标系的效果 ……………… 085
　6.4　Robot operation control　机器人动作指令 ………………………………………………086
　　6.4.1　Interpolation movement　插补指令 ………………………………………………… 086
　　6.4.2　Cnt（Continuous）连续轨迹运行 …………………………………………………… 089
　　6.4.3　Pallet operation　码垛指令 …………………………………………………………… 091
　6.5　Robot operation control　机器人运行控制 ………………………………………………094
　6.6　Detailed explanation of command words　指令详细解释 ……………………………106

Chapter 7 Functions Set with Parameters 参数设置 ········ 111

7.1 Standard tool coordinate 标准工具坐标系 ·· 111
7.2 Standard base coordinate 标准基本坐标系 ·· 114
7.3 Free plane limit 自由平面限制 ·· 118
7.4 Automatic return setting after jog feed at pause
     点动暂停后返回轨迹设置 ··· 119
7.5 Warm-up operation mode 暖机运行模式 ·· 120
7.6 High-speed RAM operation image 高速 RAM 运行图 ······························ 121
7.7 Collision detection function 碰撞检测 ··· 122
7.8 Simulation and robot calibration operation 模拟及机器人校准运行 ········· 123

Chapter 8 Advanced Functions 高级功能 ····················· 124

8.1 Configuration flag 结构标志 ··· 124
　8.1.1 Configuration flag of 6-axis robot 6 轴机器人的结构标志 ················· 124
　8.1.2 Configuration flag for horizontal multi-joint robot 水平多关节型机器人的结构标志 ··· 126
8.2 Spline interpolation 样条曲线插补 ·· 126
8.3 Ex-T control 以外部原点为基准的控制 ·· 130
　8.3.1 Outline 概述 ·· 130
　8.3.2 Ex-T coordinate setting Ex-T 坐标系设置 ······································· 131
　8.3.3 Operation 操作 ·· 133
8.4 Cooperative operation function 联合操作功能 ······································· 141

Chapter 9 Tracking Control 跟踪控制 ·························· 147

9.1 Tracking systems 跟踪系统 ·· 147
　9.1.1 Configuration example of conveyor tracking systems 传输线跟踪系统 ····· 147
　9.1.2 Configuration example of vision tracking systems 视觉跟踪系统 ······ 148
　9.1.3 Measures against the noise 抗电磁干扰 ·· 149
　9.1.4 Calibration operation for conveyor and robot 传送带和机器人的校准操作 ··· 150
　9.1.5 Tracking check function 跟踪核查功能 ··· 150
9.2 Circular arc tracking 圆弧跟踪 ··· 150

Chapter 10 Additional Axis 附加轴 ···························· 153

10.1 Outline 概述 ·· 153

10.2　Additional axis function　附加轴功能 ··· 161

Chapter 11　Maintenance and Inspection　维护和保养 ·········· 162

11.1　Robot arm structure　机器人构造 ·· 162
　11.1.1　Structure for horizontal multi-joint type robot　水平型机器人的构造 ········· 162
　11.1.2　Structure for 6-axis robot　6轴机器人的构造 ································· 163
11.2　Installing/removing the cover　盖板安装与拆卸 ····································· 164
11.3　Inspection，maintenance and replacement of timing belt
　　　同步带检查保养及张紧度调整 ··· 166
　11.3.1　Inspection and maintenance of timing belt for horizontal multi-joint type robot
　　　　　水平型机器人同步带的检测和维护 ··· 166
　11.3.2　Inspection and maintenance of timing belt for 6-axis robot
　　　　　6轴机器人同步带保养维护 ··· 167
　11.3.3　Tension adjustment of timing belt　同步带的张紧调整 ······················· 169
11.4　Lubrication　润滑 ··· 170
11.5　Replacing the battery　更换电池 ·· 172
11.6　The check of the filter，cleaning，exchange
　　　过滤窗的检查、清洗及更换 ··· 173
11.7　Overhaul　大修 ··· 175

Chapter 12　The Application of Robots in Welding Industries
　　　机器人在焊接行业中的应用 ·············· 176

12.1　Composition and structure　组成结构 ··· 177
12.2　Feature　特点 ·· 178
12.3　Structure design　结构设计 ··· 179
12.4　Equipment　装备 ··· 180
12.5　Welding application　焊接应用 ··· 180
　12.5.1　Workstation　工作站 ··· 180
　12.5.2　Box welding robot workstation　箱体焊接机器人工作站 ···················· 181
　12.5.3　Robot flexible laser welding machine　机器人柔性激光焊接机 ············· 183
　12.5.4　Shaft welding robot workstation　轴类焊接机器人工作站 ··················· 184
　12.5.5　Robot welding stud workstation　机器人焊接螺柱工作站 ··················· 184
12.6　Welding robot production line　焊接机器人生产线 ······························· 185
12.7　Application of welding robot in automobile production
　　　焊接机器人在汽车生产中的应用 ··· 187
12.8　Characteristics of arc welding　弧焊特点 ·· 188
　12.8.1　Basic function　基本功能 ·· 188

12.8.2　Welding equipment　焊接设备 ·· 189
12.8.3　Maintaining　维护保养 ·· 190

Chapter 13　The Application of Robots in Other Industries 机器人在其他行业中的应用 ································ 191

13.1　Application of robot in transportation industry 机器人在搬运码垛行业中的应用 ·· 191
13.2　Application of robot in polishing industry 机器人在打磨抛光行业中的应用 ·· 194
13.3　Application of robot in machine tool industry 机器人在机床加工行业中的应用 ·· 197
　　13.3.1　Overview　概述 ·· 197
　　13.3.2　Loading and unloading robot features　上下料机器人的特点 ················· 198
　　13.3.3　Robot classification　机器人分类 ·· 199
　　13.3.4　Working example　工作样例 ·· 202
13.4　Application of robot in cutting industry　机器人在切割行业中的应用 ······ 209
13.5　Application of robot in spraying industry　机器人在喷涂行业中的应用 ··· 210

Chapter 14　Robot Force Sensing Control　机器人的力觉控制　214

14.1　What is the force sensing control function?　什么是力觉控制功能? ···· 214
14.2　Terms　术语 ·· 215
14.3　System composition　系统构成 ·· 215
14.4　Force sense coordinate system　力觉坐标系 ··· 217
　　14.4.1　The definition of the force sense coordinate system　力觉坐标系的定义 ···· 217
　　14.4.2　Force sensor coordinate system　力觉传感器坐标系 ····························· 219
14.5　Program　程序 ·· 221
　　14.5.1　Mode switching　模式切换 ··· 221
　　14.5.2　Sample program 1　样例程序 1 ··· 221
　　14.5.3　Sample program 2　样例程序 2 ··· 224
　　14.5.4　Sample program 3　样例程序 3 ··· 225
　　14.5.5　Sample program 4　样例程序 4 ··· 228
14.6　Stiffness controlling　刚度控制 ··· 229
　　14.6.1　Stiffness controlling principle　刚度控制原理 ··· 229
　　14.6.2　Change control characteristics 1　改变控制特性 1 ································· 231
　　14.6.3　Change control characteristics 2　改变控制特性 2 ································· 231
14.7　Assembly　装配 ·· 234
　　14.7.1　Case 1　案例 1 ·· 234
　　14.7.2　Case 2　案例 2 ·· 235

14.7.3　Case 3: data latching and reading　案例3：数据锁存及读取 ·················· 237
14.7.4　Data transfer case　数据传送案例 ··· 238
14.7.5　Position alignment push in　位置对准推入 ·· 238

References　参考文献 ··· 241

Chapter 1
Standard Specifications of the Robot
机器人的技术规格

1.1　Robot arm　机器人本体

（1）Parts names of the 6-axis robot　6 轴机器人各部分的名称（图 1-1）

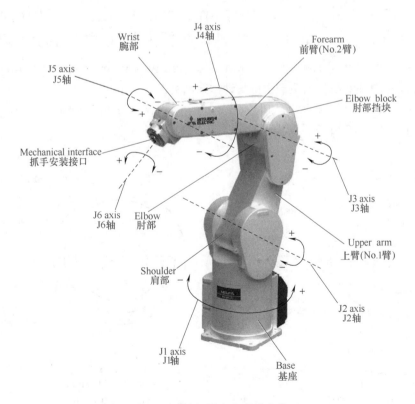

图 1-1　6 轴机器人各部分名称

（2）Parts names of the 4-axis robot 4轴机器人各部分名称（图1-2）

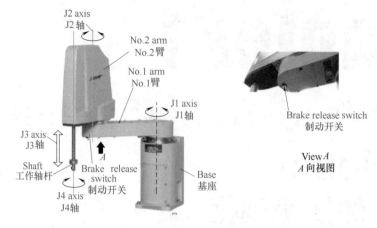

图1-2 4轴机器人各部分名称

1.2 Rated load and operating range 额定负载与动作范围

1.2.1 Relationship between mass capacity and movement area for 4-axis robot 4轴机器人负载质量与动作范围关系

（1）Relationship between the mass capacity and movement area for 4-axis robot（6kg） 4轴6kg机器人负载质量与动作范围关系（图1-3）

（2）Relationship between the mass capacity and movement area for 4-axis robot（12kg） 4轴12kg机器人负载质量与动作范围关系（图1-4）

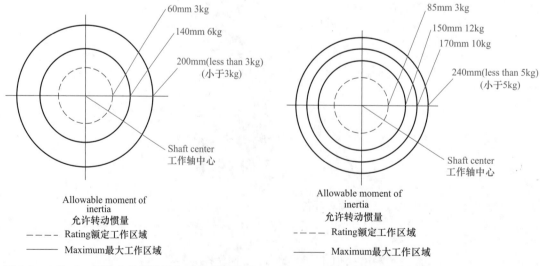

图1-3 4轴6kg机器人负载质量与动作范围关系　　图1-4 4轴12kg机器人负载质量与动作范围关系

1.2.2 Relationship between mass capacity and movement area for 6-axis robot 6轴机器人负载质量与动作范围关系

以 6 轴机器人的 J5 轴旋转中心为纵坐标，J6 轴旋转中心为横坐标，在不同的范围内，机器人可抓取的工作负载质量不同。

（1）Relationship between mass capacity and movement area for 6-axis robot (7kg) 6轴 7kg 机器人负载质量与动作范围关系（图 1-5）

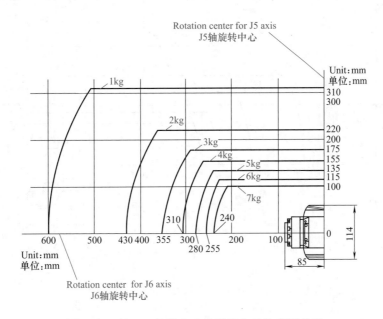

图 1-5 6 轴 7kg 机器人负载质量与动作范围关系

（2）Relationship between mass capacity and movement area for 6-axis robot (13kg) 6轴 13kg 机器人负载质量与动作范围关系（图 1-6）

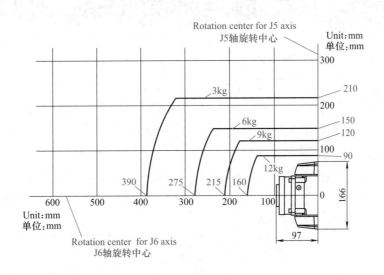

图 1-6 6 轴 13kg 负载质量与动作范围关系

（3）Relationship between mass capacity and movement area for 6-axis robot (20kg) 6轴20kg机器人负载质量与动作范围关系（图1-7）

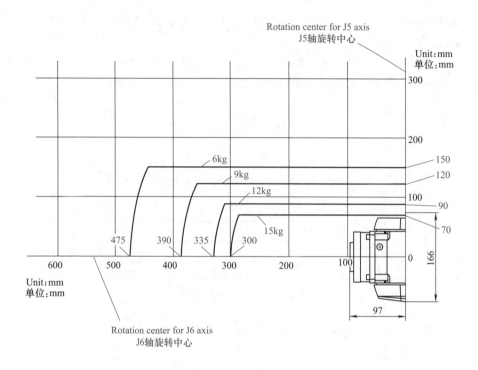

图1-7　6轴20kg机器人负载质量与动作范围关系

1.3　Relationship between mass capacity and speed　负载质量与速度的关系（图1-8）

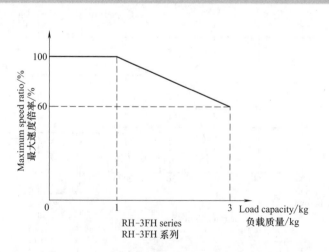

图1-8　负载质量与速度的关系

1.4 Relationship between height of shaft（J3 axis）and acceleration/deceleration speed J3 轴行程高度与加减速度的关系

（1）Area in which acceleration/deceleration speed is compensated　加减速速度自动补偿区（图 1-9）

Area 1——Area in which speed and acceleration/deceleration speed are not compensated.

在区间 1，不能对速度和加减速度进行补偿。

Area 2——Area in which speed and acceleration/deceleration speed are compensated.

在区间 2，可以对速度和加减速度进行补偿。

（2）Recommended path when positioning at the bottom edge of the Z axis　定位点在 Z 轴底部时的推荐路径

Time to reach the position precision：

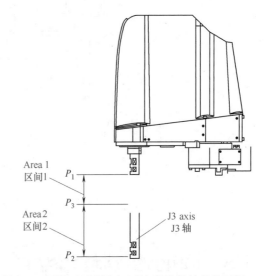

图 1-9　可以对速度和加减速度进行补偿的区间

When using this robot, the time to reach the position repeatability may be prolonged due to the effect of residual vibration at the time of stopping. If this happens, take the following measures.

1）Change the operation position of the Z axis to the location near the top as much as possible.

2）Increase the operation speed prior to stopping.

3）When positioning the work near the bottom edge of the Z axis, if no effectiveness is achieved in step "2" above, perform operation path 1（robot path: $O \to A \to C$）. In the case of operation path 2（robot path: $O \to B \to C$）, residual vibration may occur（refer to Fig. 1-10）.

到达位置精度内的时间：

在使用机器人时，由于停止时残留振动的影响，达到位置精度内的时间有可能变长，此时应执行以下处理。

1）将 Z 轴的停止位置尽量改为上部位置。

2）提高停止前的动作速度。

3）在 Z 轴最下端附近进行定位（如图 1-10 路径 2）有振动时，应执行路径 1 的动作。

路径 1：$O \to A \to C$。
路径 2：$O \to B \to C$。
按路径 2 运行有时会发生残留振动（参阅图 1-10）。

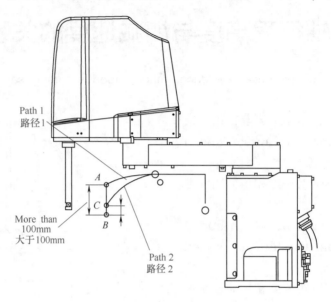

图 1-10　定位点在 Z 轴底部时的推荐路径

1.5　Outside dimensions　外形尺寸（图 1-11、图 1-12）

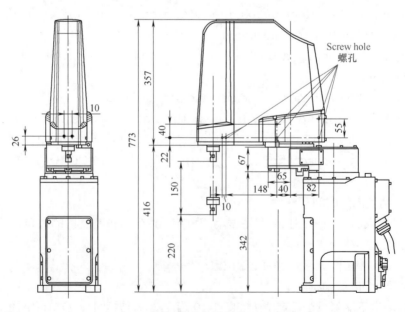

图 1-11　外形尺寸

Chapter 1　Standard Specifications of the Robot

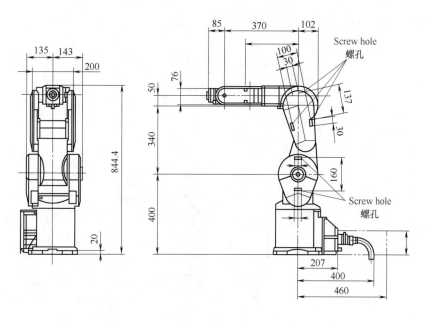

图 1-12　6 轴机器人（RV-7F）外形尺寸

1.6　Tooling　工具（外围附件）

（1）Wiring and piping for hand　抓手的线路与管道（图 1-13）

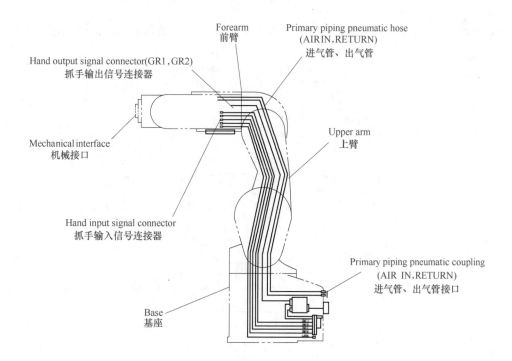

图 1-13　机器人本体内的线路管路

（2）Pipeline arrangement　管路布置（图 1-14）

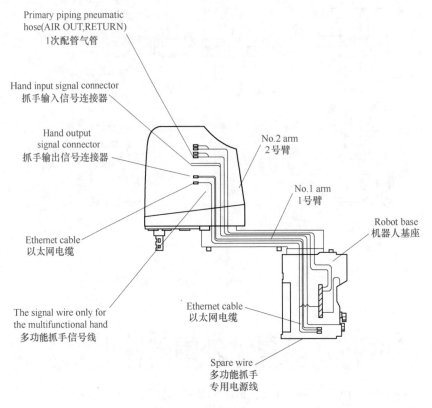

图 1-14　管路结构

（3）Location of screw holes for fixing wiring/piping　固定线路及管路的螺钉孔（图 1-15）

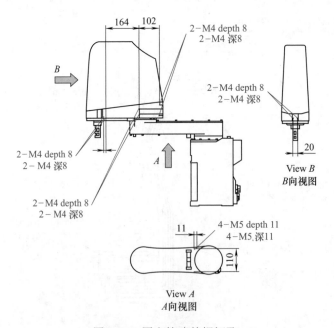

图 1-15　固定管路的螺钉孔

Chapter 1　Standard Specifications of the Robot

（4）Example of wiring and piping 1　线路及管路连接样例1（图1-16）

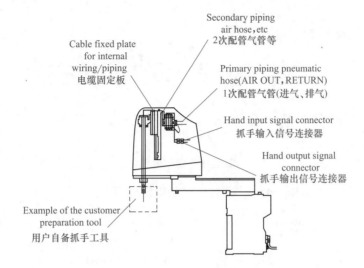

图1-16　管路连接样例

（5）Example of wiring and piping 2　线路及管路连接样例2（图1-17）

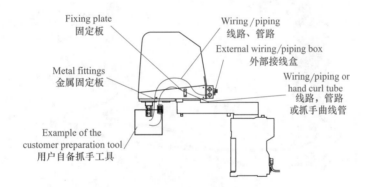

图1-17　线路及管路连接样例

（6）Installation of the air pipe　通气管的安装（图1-18）

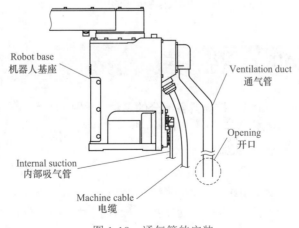

图1-18　通气管的安装

(7) Air supply circuit example for the hand　供气管路（图1-19）

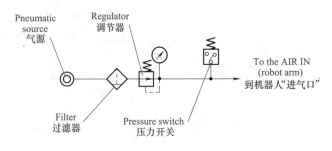

图1-19　供气管路

(8) Fixing of the flexible cable　柔性电缆的固定（图1-20）

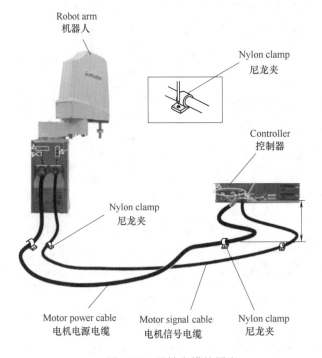

图1-20　柔性电缆的固定

1.7　Controller　控制器

(1) Names of each part 1　控制器各部分的名称1（图1-21）

① Earth leakage breaker——Connect the primary power source.

漏电保护断路器——连接主电源。

② Grounding plate——The grounding terminal for grounding the cable. Strip off the sheath of the cable and ground the controller case using this plate.

接地板——安装接地电缆的端子。使用接地板连接剥去绝缘外层的电缆，用于控制器接地。

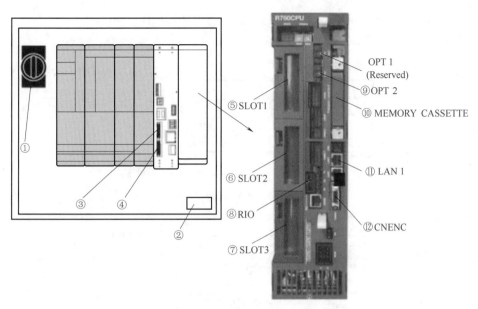

图 1-21 控制器各部分的名称 1

③ EMG1 connector——External emergency stop input, door switch input, enabling device switch, and magnet contactor control connector output for addition axes.

急停 1 接口——用于外部急停，门开关信号输入，使能开关信号输入，附加轴接触器控制信号。

④ EMG2 connector——Emergency stop output, mode output, robot error output, and special stop input (SKIP).

急停 2 接口——急停输出、模式信号输出、机器人故障信号输出、特殊停止输入。

⑤ ~ ⑦ Option slot (SLOT1, SLOT2, SLOT3)——Install the interface optional.

选件插槽——安装选件的插槽。

⑧ Extension parallel input/output unit connection connector (RIO)——Connect the extension parallel input/output unit.

扩展 I/O 卡接口。

⑨ Addition axis connection connector (OPT2)——Connect the cable for addition axis control.

附加轴接口——连接附加轴控制器电缆。

⑩ Expansion memory cassette (MEMORY CASSETTE)——Install the memory cassette optional.

扩展存储卡插槽——安装存储卡。

⑪ Ethernet interface (LAN1)——Connect the Ethernet cable.

以太网接口——连接以太网电缆。

⑫ Tracking interface——Connect the encoder cable, if it uses the tracking function.

CNENC 追踪功能接口——使用追踪功能时，连接编码器电缆。

（2）Names of each part 2　控制器各部分的名称 2（图 1-22）

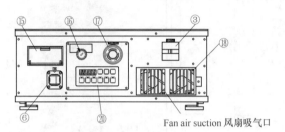

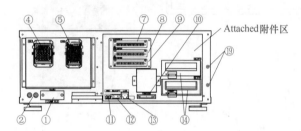

图 1-22　控制器各部分的名称 2

① ACIN terminal——The terminal box for AC power source（single phase，AC200V）input.

ACIN 连接器——AC 电源（单相，AC200V）输入用端子盒。

② PE terminal——The screw for grounding of the cable（M4 screw×2 places）.

接地电缆用螺栓。

③ Power switch——This turns the control power ON/OFF（with earth leakage breaker function）.

电源 ON/OFF 开关（带漏电保护功能）。

④ Machine cable connector（motor signal）（CN1）——Connect with the CN1 connector of the robot arm.

电机信号电缆插口（CN1）——与机器人本体的 CN1 口连接。

⑤ Machine cable connector（motor power）（CN2）——Connect with the CN2 connector of the robot arm.

电机电源电缆插口（CN2）——与机器人本体的 CN2 口连接。

⑥ T/B connection connector（TB）——This is a dedicated connector for connecting the T/B. When not using T/B，connect the attached dummy connector.

手持单元 T/B 接口——这是手持单元 T/B 的专用接口。如果不使用手持单元 T/B 时，要连接一个空插头。

⑦~⑩ CNUSR connector——The connector for input/output connection dedicated for robot（a plug connector attached）.

机器人专用的输入输出卡插口。

⑪ LAN connector（LAN）——For LAN connection.

LAN 网线插口。

Chapter 1 Standard Specifications of the Robot

⑫ ExtOPT connector（ExtOPT）——Connect the cable for addition axis control.
ExtOPT 是连接附加轴电缆的插口。

⑬ RIO connector（RIO）——Connect the extension parallel input/output unit.
远程输入输出卡插口——用于连接扩展的输入输出单元。

⑭ Option slot（SLOT1，SLOT2）——Install the interface optional（install the cover, when not using）.
选件插口——用于安装选件卡（不使用时需要安装盖子）。

⑮ Interface cover——USB interface and battery are mounted.
USB 插口及电池安装盒。

⑯ Mode key switch——This key switch changes the robot's operation mode.
模式选择键——用于转换机器人工作模式。

⑰ Emergency stop switch——This switch stops the robot in an emergency state. The servo turns OFF.
急停开关——这个开关用于使机器人进入急停状态，同时使伺服系统 OFF。

⑱ Filter cover——There is an air filter inside the cover.
过滤器盖板——盖板之下有一空气过滤器。

⑲ Grounding terminal——The grounding terminal for connecting cables of option card（M3 screw × 2 pieces）.
接地端子——这是用于连接接地电缆的接线端子。

⑳ Operation panel——The operation panel for servo ON/OFF, START/STOP the program, etc.
操作面板——这是用于执行伺服 ON/ 伺服 OFF、程序启动 / 停止等操作的面板。
以下内容为操作按键等，图中略。

㉑ Display panel（STATUS. NUMBER）——The alarm number, program number, override value（%）, etc., are displayed.
显示面板（状态，数字）——显示报警号、程序号、速度倍率等信息。

㉒ CHNGDISP button——This button changes the details displayed on the display panel in the order of "Override" → "Line No." → "Program No." → "User information." → "Maker information".
显示内容改变按键——使用本按键顺序改变显示内容："速度倍率" → "程序行号" → "程序号" → "用户信息" → "制造商信息"。

㉓ UP/DOWN button——This scrolls up or down the details displayed on the "STATUS, NUMBER" display panel.
上下翻页按键——使用本按键，以上下翻页的形式，显示 "状态、数字" 等信息。

㉔ SVO. ON button——This turns ON the servo power. The servo turns ON.
伺服 ON 操作按键——使用本按键，执行 "伺服 ON" 操作。

㉕ SVO. OFF button——This turns OFF the servo power. The servo turns OFF.
伺服 OFF 操作按键——使用本按键，执行 "伺服 OFF" 操作。

㉖ START button——This executes the program and operates the robot. The program is run continuously.

启动按键——使用本按键,执行"程序启动"操作。注意程序是连续执行。

㉗ STOP button —— This stops the robot immediately. The servo does not turn OFF.

停止按键——使用本按键,执行"程序停止"操作。注意伺服系统并不处于 OFF 状态。

㉘ RESET button——This resets the error. This also resets the program's halted state and resets the program.

复位按键——使用本按键,执行"解除故障报警"操作,同时也解除程序的停止状态,使程序复位。

㉙ END button——This stops the program being executed at the last line or END statement.

结束按键——使正在执行的程序在最后一行或 END 行停止。

(3) Installation dimensions 1　安装空间要求 1 (图 1-23)

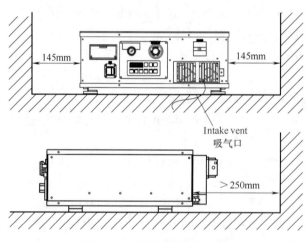

图 1-23　水平放置安装空间

(4) Installation dimensions 2　安装空间要求 2 (图 1-24)

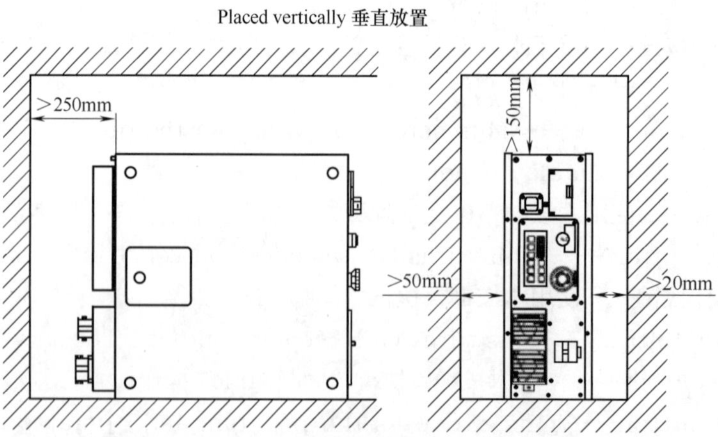

图 1-24　垂直放置安装空间

（5）Emergency stop input and output　急停信号的输入输出（图1-25）

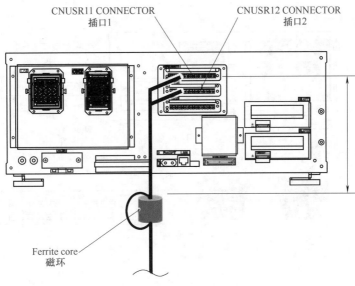

图1-25　急停信号的输入输出

（6）Method of wiring for external emergency stop connection 1　外部急停电缆的安装连接方法1（图1-26）

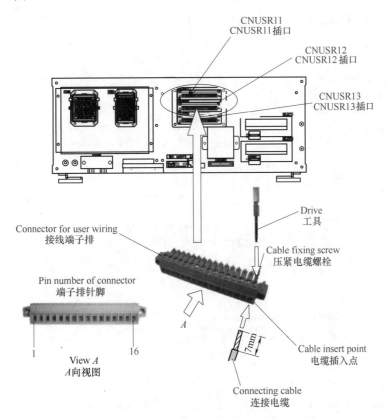

图1-26　外部急停电缆的安装连接方法

(7) Method of wiring for external emergency stop connection 2 外部急停电缆的安装连接方法2（图1-27）

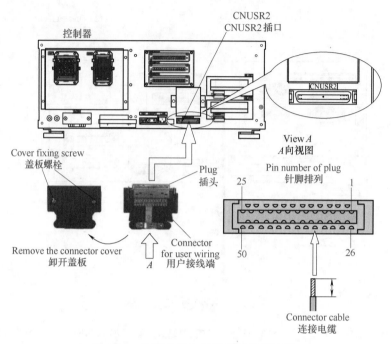

图1-27 外部急停电缆的安装连接方法

(8) Door switch function 门开关功能（图1-28）

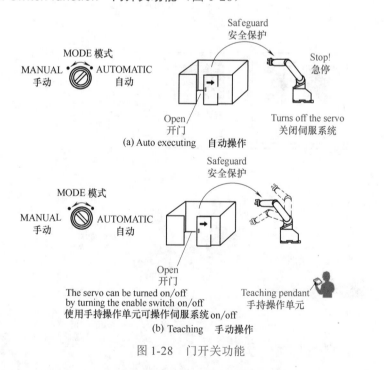

图1-28 门开关功能

(9) Mode changeover switch input 模式选择开关的使用

AUTOMATIC——The operation from external equipment becomes available.

Chapter 1 Standard Specifications of the Robot

Operation which needs the right of operation from T/B cannot be performed. It is necessary to set the parameter for the right of operation to connection with external equipment.

自动模式——通过外部设备进行的操作有效。无法进行需要示教单元操作权的操作。与外部设备的连接中，需要对操作权用的参数进行设置。

MANUAL——When T/B is available, only the operation from T/B becomes available. Operation which needs the right of operation from external equipment cannot be performed.

手动模式——仅通过示教单元进行的操作有效。无法进行需要外部设备操作权的操作。

如图1-29。

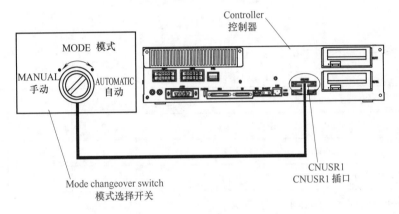

图1-29 模式选择开关

（10）Connection of the mode changeover switch input 模式选择开关输入信号的连接（图1-30）

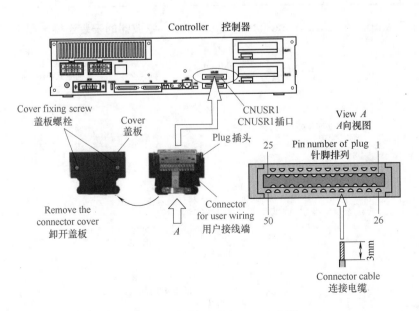

图1-30 模式选择开关输入信号的连接

1.8 Options 选件

(1) Solenoid valve set 电磁阀组(图1-31)

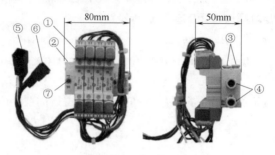

图1-31 电磁阀组

电磁阀组各零部件的名称如表1-1所示。

表1-1 电磁阀组各零部件的名称

序号	Part name 部件名称	1 set 1组	2 set 2组	3 set 3组	4 set 4组
①	Solenoid valve 电磁阀	1	2	3	4
②	Manifold 阀座	1	1	1	1
③	Quick coupling (A/B port) 接线口 A/B	8	8	8	8
④	Quick coupling (P/R port) 接线口 P/R	2	2	2	2
⑤	Connector 插头	1	1	2	2
⑥	Contact 插座	3	5	8	10
⑦	Installation screw 安装螺栓	2	2	2	2

(2) Example of EMC noise filter installation 滤波器的安装连接(图1-32)

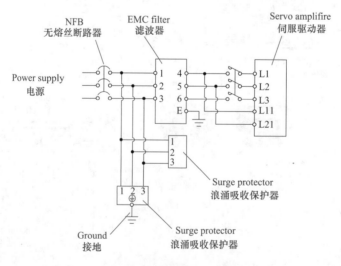

图1-32 滤波器的安装连接

Chapter 1　Standard Specifications of the Robot

（3）Teaching pendant（T/B）　示教单元的按键布置和主要功能（图 1-33）

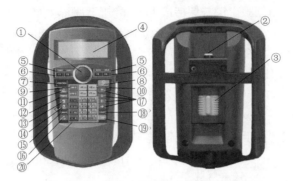

图 1-33　示教单元的按键布置和主要功能

① [Emergency stop] switch——The robot servo turns OFF and the operation stops immediately.

[急停] 开关——紧急切断伺服系统并停止操作。

② [Enable/Disable] switch——This switch changes the T/B key operation between enable and disable.

[使能切换] 开关——用于切换 T/B 单元的"使能状态"。

③ [Enable] switch——When the [Enable/Disable] switch "②" is enabled, and this key is released or pressed with force, the servo will turn OFF, and the operating robot will stop immediately.

[使能] 开关——当 [使能切换] 开关处于"使能状态",如果用力压下或释放 [使能] 开关,则机器人的伺服系统被切断,操作立即停止。

④ LCD display panel——The robot status and various menus are displayed.

LCD 显示屏——显示机器人工作状态和各菜单。

⑤ Status display lamp——Display the state of the robot or T/B.

状态显示灯——显示"机器人"或"T/B"的状态。

⑥ [F1], [F2], [F3], [F4]——Execute the function corresponding to each function currently displayed on LCD.

[F1], [F2], [F3], [F4]——选择执行在当前屏幕上对应的功能。

⑦ [FUNCTION] key——Change the function display of LCD.

[功能] 键——切换 LCD 上显示的功能。

⑧ [STOP] key——This stops the program and decelerates the robot to a stop.

[停止] 键——停止程序并使机器人动作减速停止。

⑨ [OVRD ↑] [OVRD ↓] key——Change moving speed. Speed goes up by [OVRD ↑] key. Speed goes down by [OVRD ↓] key.

[OVRD] 键——速度调整。速度增加使用 [OVRD ↑] 键,速度降低使用 [OVRD ↓] 键。

⑩ [JOG] operation key——Move the robot according to jog mode, or input the numerical value.

[JOG 模式选择] 键——选择进入 [JOG 模式],也可以输入数字量。

⑪ [SERVO] key——Press this key with holding AA key lightly, then servo power will turn on.

[伺服]键——按下[伺服]键并保持按下 AA 键，伺服 =ON。

⑫ [MONITOR] key——It becomes monitor mode and display the monitor menu.

[监视]键——使用[监视]键，可以使系统进入"监视模式"并显示"监视菜单"。

⑬ [JOG] key——It becomes jog mode and display the jog operation.

[JOG]键——进入"点动模式"并显示点动操作。

⑭ [HAND] key——It becomes hand mode and display the hand operation.

[抓手]键——进入"抓手模式"并显示抓手操作。

⑮ [CHARCTER] key—— This changes the edit screen, and changes between numbers and alphabetic characters.

[字符]键——切换编辑屏幕，切换数字输入和字符输入。

⑯ [RESET] key—— This resets the error. The program reset will execute, if this key and the EXE key are pressed.

[复位]键——解除故障报警信息。与[EXE]键同时使用，可使程序复位。

⑰ [↑][↓][←][→] key——Moves the cursor each direction.

[光标]键——在各个方向移动光标。

⑱ [CLEAR] key —— Erase the one character on the cursor position.

[清除]键——擦除在光标位置的字符。

⑲ [EXE] key—— Input operation is fixed. While pressing this key, the robot moves when direct mode.

[执行]键——应用发出操作执行指令。在按下[EXE]键，机器人进入直接动作模式时，机器人动作。

⑳ Number/Character key—— Erase the one character on the cursor position. And, inputs the number or character.

[数字/字符]键——擦除在光标位置上的字符，同时输入数字或字符。

（4）Parallel I/O interface　I/O 卡（图 1-34）

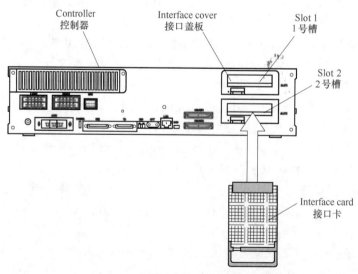

图 1-34　I/O 卡

Chapter 1　Standard Specifications of the Robot

（5）Connect of power supply　电源连接（图 1-35）

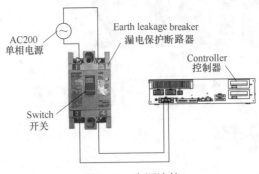

图 1-35　电源连接

（6）Connect of parallel I/O unit　I/O 单元的连接（图 1-36）

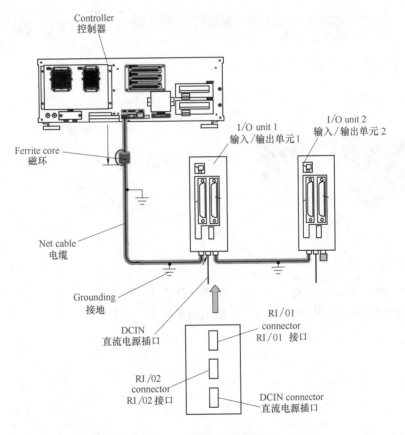

图 1-36　I/O 单元的连接样例

Chapter 2
Installing the Robot
机器人的安装

2.1　Unpacking to installation　开箱及安装

（1）Unpacking　开箱（图 2-1）

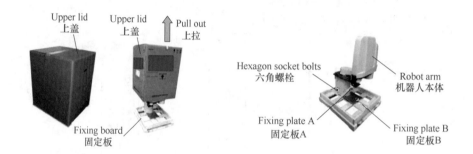

图 2-1　开箱

（2）Transportation of robot arm 1　搬运机器人 1（图 2-2）
（3）Transportation of robot arm 2　搬运机器人 2（图 2-3）

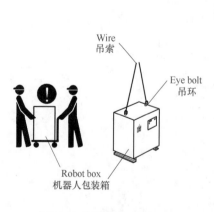

图 2-2　搬运机器人 1

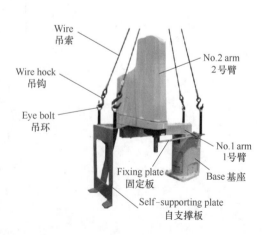

图 2-3　搬运机器人 2

（4）Transporting with a crane robot 1　吊装机器人1（图2-4）
（5）Transporting with a crane robot 2　吊装机器人2（图2-5）

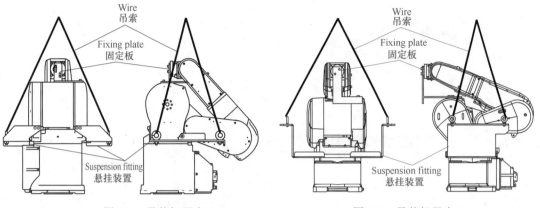

图 2-4　吊装机器人 1　　　　　图 2-5　吊装机器人 2

2.2 Connecting the power cable and grounding cable
连接电源电缆和接地电缆

（1）Connecting the power cable and grounding cable　连接电源电缆和接地电缆（图2-6）

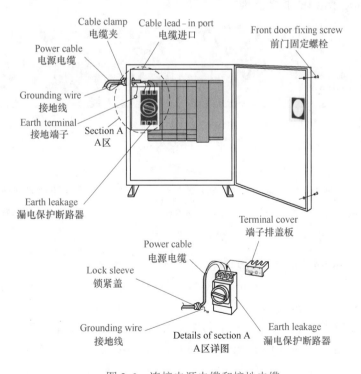

图 2-6　连接电源电缆和接地电缆

（2）Connecting the power supply　电源的连接（图2-7）

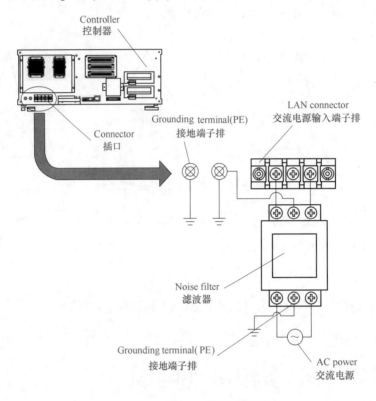

图2-7　电源的连接

（3）Grounding methods　接地方式（图2-8）

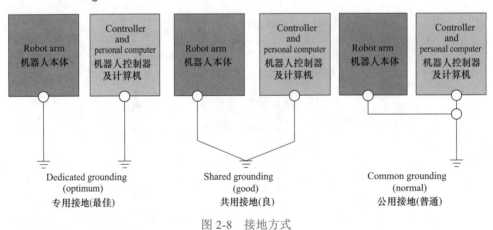

图2-8　接地方式

（4）Grounding procedures　接地规程

1）Prepare the grounding cable (AWG #11/4.2mm^2 or more) and robot side installation screw and washer.

准备接地用电缆（AWG #11/ 4.2mm^2 以上）及机器人侧的安装螺栓及垫圈。

2）If there is rust or paint on the grounding screw section (A), remove it with a file, etc.

接地螺栓部位（A）如有锈或油漆，应使用锉刀等去除。
3）Connect the grounding cable to the grounding screw section.
将接地电缆连接到接地螺栓部位。
见图 2-9。

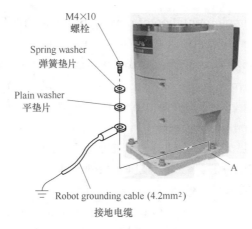

图 2-9　连接接地电缆

(5) Connecting the machine cables　连接控制器（图 2-10）

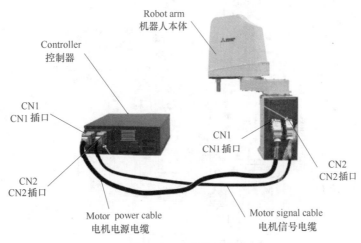

图 2-10　连接控制器

(6) Connecting the Ethernet cable　连接以太网电缆（图 2-11）

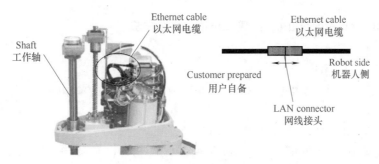

图 2-11　连接以太网电缆

2.3 Attachments installation procedures
附件安装

(1) Precautions for implementation of moving part 可动部分实装注意事项（图 2-12）

1) The internal air hoses and cables are bent and twisted according to the operations of the J3 and J4 axes. Structurally, the twisting occurs inside the shaft, and the bending occurs inside the expanding sleeve. Without the expanding sleeve, the twisted air hoses and cables may run onto the guide portion of the fixing plate, and the air hoses may be bent or broken.

内装的气管、电缆根据 J3 轴、J4 轴的动作进行弯曲和扭曲。结构设置为：在轴内部扭曲，在扩展套管部弯曲。不使用扩展套管时，在固定板的导向部分可能会发生扭曲的气管和电缆交错，从而导致气管折断或电缆断线。

2) Do not place the connector connecting part and the air hose relay part in the binding or twisting range.

请勿将插头连接部和气管的中继部设置在弯曲范围及扭曲范围内。

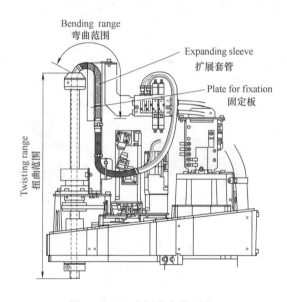

图 2-12　运动部件安装要点

(2) Installation procedure 气管及电缆的内装方法

The installation procedure is shown in Fig. 2-13.

安装方法如图 2-13。

1) Move the J3 axis to the top end with a jog operation and shut off the controller's power supply.

使用 JOG 操作使 J3 轴移动至上端，并切断控制器的电源。

2) Remove the screws fixing the No. 2 arm cover U, and remove the No. 2 arm cover U.

Chapter 2　Installing the Robot

卸下固定 2 号机械臂盖板 U 的安装螺栓，卸下 2 号机械臂盖板 U。

3) Pass the tool (hand) side of the internal air hoses and cables through the shaft. Make sure that the air hoses and the cables are not twisted or crossed.

使 tool（抓手）的内装气管与电缆穿过轴内。确保气管、电缆无绞扭、交叉。

4) Fix the air hoses and the cables with a cable tie so as to position the end of the expanding sleeve 10mm away from the opening of the shaft.

使用扎带固定套管，使扩展套管上端在距离轴开口部约 10mm 位置。

5) The highest point of the curved section of the air hoses and cables should be matched up with the top of the fixing plate.

将气管和电缆的弯曲部顶点设为与固定板上端同一高度。

6) In the state of the steps 4) and 5), fix the air hoses and the cables to the "a" and "b" portions of the plate with cable ties.

在步骤 4)、5) 的状态下，将气管、电缆扎紧在固定板的"a""b"部。

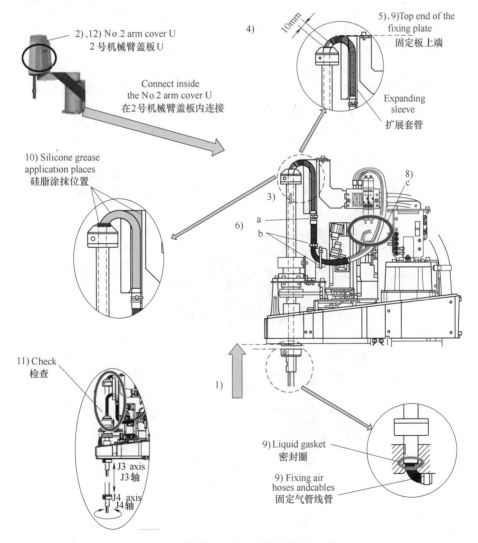

图 2-13　安装机器人内部气管和电缆

7) Installation of the hand input signal or the optional solenoid valve.
安装抓手输入信号或电磁阀组。

8) The connected connectors are stored to the "c" portion.
将连接器收至"c"部。

9) Carry out piping and wiring on the tool side.
安装 tool 侧的配管、配线。

10) Apply silicon grease to the contact surface between the air hoses of the fixing plate and the cables, the cable sliding portion from the shaft upper end to the fixed portion, and the opening on the shaft upper end.
在固定板的气管与电缆的接触面、轴上端部与固定部位之间的电缆转动部、轴上端部的开口处涂抹硅脂。

11) Power on the controller, perform the jog operation for the J3 and J4 axes, and check that the air hoses and the cables do not interfere with other components.
接通控制器的电源，使用 JOG 运行使 J3 轴、J4 轴动作，以确认气管及电缆与其他零部件不发生干涉。

12) Turn off the controller's power supply, then install the No. 2 arm cover U securely as before with fixing screws.
切断控制器的电源后，用固定螺栓将 2 号机械臂盖板 U 按原样牢固安装。

(3) Installing the solenoid valve set　安装电磁阀组（图 2-14）

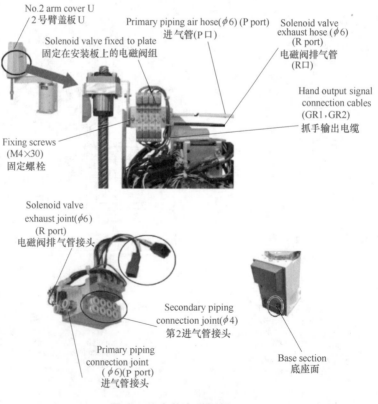

图 2-14　安装电磁阀组

（4）Installing the forearm external wiring set　安装前臂外部电缆组（图 2-15）

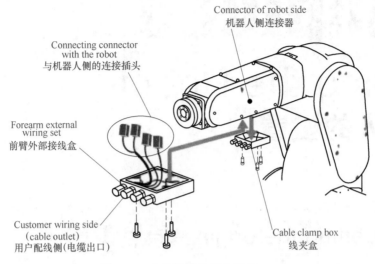

图 2-15　安装前臂外部电缆组

（5）Installing the Base external wiring set　安装基座外部电缆组（图 2-16）

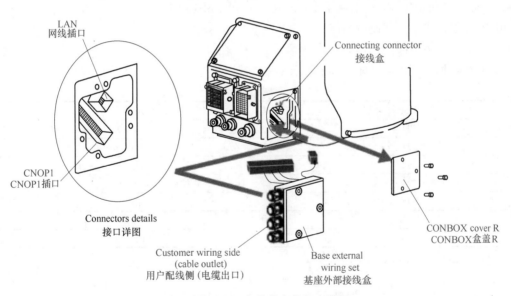

图 2-16　安装基座外部电缆组

Chapter 3
Setting the Robot
机器人的设置

3.1　Setting the origin　原点设置

3.1.1　Origin data input method　数据输入方式

Setting the origin with the origin data input method.
使用原点数据输入方法设置原点。

1）Confirming the origin data.
确认原点数据。原点数据如表 3-1 所示（由制造商给出）。

表 3-1　原点数据设置记录表

Data	Default 出厂值
D	V!#S29
J1	06DTYY
J2	2?HL9X
J3	1CP55V
J4	T6!M$Y
J5	Z21J%Z
J6	A12%Z0

2）Turning on the control power.
上电。

3）Set the mode of the controller to "MANUAL".
操作设置控制器进入"手动模式"（图 3-1）。

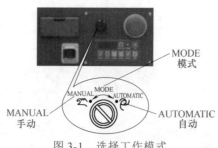

图 3-1　选择工作模式

3.1.2 Preparing the T/B　操作使用 T/B 示教单元（图 3-2）

Set the T/B [ENABLE] switch to "ENABLE". The menu selection screen will appear. The following operations are carried out with the T/B.

将 T/B 示教单元的 [ENABLE] 开关设置为"ENABLE（使能有效）"状态。显示屏上出现"菜单"画面，使用 T/B 单元进行以下操作。

Selecting the origin setting method.

选择原点设置方式。

1) Press the [4] key on the menu screen, and display the origin/brake screen.

在"菜单"界面按下 [4] 键，显示"ORIGIN/BRAKE"界面（"ORIGIN"在界面中显示缩写"ORIGN"）。如图 3-3。

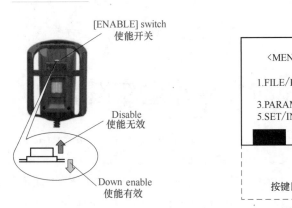

图 3-2　操作"使能开关"

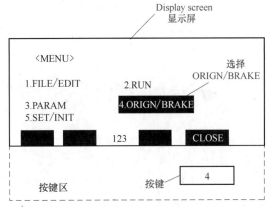

图 3-3　选择原点或 BRAKE 设置模式

2) Press the [1] key on the origin/brake screen, and display the origin setting method selection screen.

在"ORIGIN/BRAKE"界面中按下 [1] 键，显示"原点设置方式选择"界面（图 3-4）。

3) Press the [1] key on the origin setting method selection screen, and select the data input method.

在"原点设置方式选择"画面中按下 [1] 键，选择"数据输入方式"（图 3-5）。

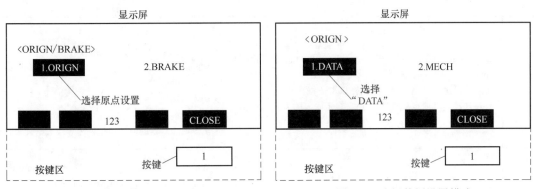

图 3-4　选择原点设置模式　　　　图 3-5　选择数据设置模式

4) Display the origin data input screen. Input the data in Table 3-1. The setting is done.

显示"原点数据输入"画面。将表 3-1 中的数据写入"数据设置"界面（图 3-6）。至此，原点设置完成（图 3-7）。

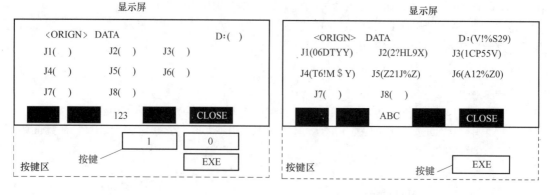

图 3-6　数据设置界面　　　　　　　图 3-7　数据设置完成界面

3.2　Resetting the origin　原点重新设置

　　The origin is set so that the robot can be used with a high accuracy. After purchasing the robot, always carry out this step before starting work. The origin must be reset if the combination of robot and controller being used is changed or if the motor is changed causing an encoder area. The origin setting methods and when each origin setting method required are shown in Table 3-2.

　　设定原点是为了能高精度地使用机器人而进行的操作。在机器人开始工作之前必须进行"原点设置"。在（使用之后），如果发生控制器与机器人的连接脱开、更换电机、编码器出现故障等情况，要进行"重新设置原点"。"重新设置原点"的方式类型和各设置方式的应用场合如表 3-2 所示。

表 3-2　原点重新设置方式

No. 序号	Method 方式	Explanation 说明	Cases when setting the origin is required 应用场合
1	Origin data input method 原点数据输入方式	The origin data set as the default is input from the T/B. Use this method at the initial startup 原点数据输入方式是将出厂时设定的原点数据通过手持单元输入的方式。在初次启动机器人使用这种方式	• At the initial startup • When the controller is replaced • When the data is lost due to flat battery of the robot controller • 初次启动时 • 重新安装控制器时 • 机器人控制器的电池电量耗尽导致数据丢失时
2	Jig method 校正棒方式	The origin posture is set with the calibration jig installed 校正棒方式是使用校正工具对原点进行设定	• When a structural part of the robot（motor, reduction gear, timing belt, etc）is replaced • When deviation occurred by a collision. • 更换机器人部件（电机、减速机、同步带等）时 • 由于碰撞等导致部件之间发生了偏移时

续表

No. 序号	Method 方式	Explanation 说明	Cases when setting the origin is required 应用场合
3	Mechanical stopper method 机械挡块方式	This origin posture is set by contacting each axis against the mechanical stopper 机械挡块方式使用各轴的机械挡块进行原点设置	• When a structural part of the robot (motor, reduction gear, timing belt, etc.) is replaced • When deviation occurred by a collision • 更换机器人部件（电机、减速机、同步带等）时 • 由于碰撞等导致部件之间发生了偏移时
4	ABS origin method ABS 原点设置方式	This method is used when the encoder backup data lost in the cause such as battery cutting 由于电池电量耗尽等原因导致编码器备份数据丢失时，使用 ABS 原点设置方式进行原点设定	• When the encoder data is lost due to flat battery of the robot arm • 在机器人本体的电池电量耗尽导致编码器数据丢失时
5	User origin method 用户原点设置方式	A randomly designated position is set as the origin posture 用户原点设置方式是将任意位置设置为原点	• When an arbitrary position is set as the origin • 需要设置"任意位置"为原点时

Caution 注意：

• The origin is set using the jig method (No. 2) at factory default.

机器人出厂时，已通过校正棒方式（No. 2）进行了原点设定。

• The value set with the jig method is encoded and used as the origin data to be input at the initial startup after shipment. When the robot arm does not mechanically deviate (for example caused by replacement of the reduction gear, motor, or timing belt) or does not lose the encoder data, the origin data input method at shipment can be used to set the origin.

使用校正棒方式获得的数据被封装起来作为原点数据。在出厂后的初次使用时输入"原点数据"。当机器人本体没有发生机械拆分（如更换减速齿轮、电机、同步带）或丢失编码器数据，出厂设置的原点数据就可以用于设置"原点数据"。

• The origin data is inherent to the serial number of each robot arm.

每一序列号的机器人其原点数据是独有的。

• The ABS origin method is used to restore the previous data by aligning the triangular marks to each other for each axis to set the lost origin data.

ABS 原点设置方式是使用各轴对准轴上的三角形标志的方法，重新恢复各轴原有的原点数据。

Caution 注意：

• The ABS origin method cannot be used when the robot arm mechanically deviates (for example caused by replacement of the reduction gear, motor, or timing belt).

在发生机器人本体的机械性拆分时（更换减速齿轮、电机、同步带）不能使用 ABS 原点设置方法。

• After the origin setting is completed, move the robot arm to the position where the ABS marks align each other, and check that the displayed joint coordinates of the position are correct.

原点设定完成后,务必将机器人本体移动至各轴ABS标记位置,并在显示屏上确认该位置的关节坐标是否显示正确。

3.2.1　Mechanical stopper method　机械挡块方式

Setting for 4-axis robot.　4轴机器人的设置。

(1) J1 axis origin setting (mechanical stopper)　J1轴原点设置

1) Press the [4] key on the menu screen, and display the origin/brake selection screen.

在"菜单"界面,按下键盘区的[4]键,选择"原点/制动器"界面(图3-8)。

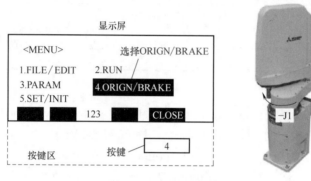

图3-8　J1轴原点设置

2) With both hands, slowly move the J1 axis in "-"(minus)direction, and contact the axis against the mechanical stopper.

用双手将J1轴缓慢地向"-"(负)方向移动,直至碰到"机械限位器"。

3) Press the [1] key, and display the Origin setting selection screen.

按下[1]键,选择原点设置选择界面(图3-9)。

4) Press the [2] key, and display the mechanical stopper selection screen.

按下[2]键选择机械限位器方式(图3-10)。

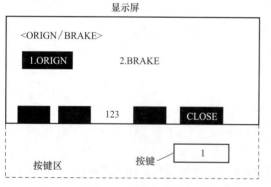

图3-9　J1轴原点设置选择

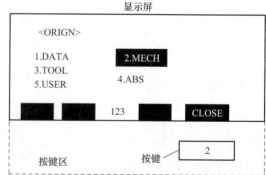

图3-10　J1轴原点设置(选择机械挡块方式)

5) Input "1" into the J1 axis. Set "0" to other axes.

按下[↑]~[→]键，将光标移至J1的"( )"内，按下[1]键。在其他的轴中设置[0]（图3-11）。

6) Press the [EXE] key, and display confirmation screen.

按下[EXE]键，接着显示确认画面（图3-12）。

7) Press the [F1] key, and the origin position is set up.

按下[F1]键，确认设置原点。

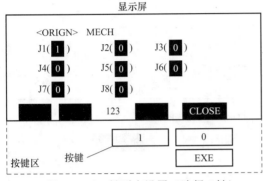

图3-11 J1轴原点设置（选择J轴）　　图3-12 J1轴原点设置（确认设置）

8) Setting of the origin is completed.

原点设置完成（图3-13）。

9) Record the origin data on the origin data seal.

将原点数据记录到原点数据表中。

(2) J2 axis origin setting (mechanical stopper) J2轴原点设置

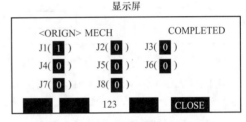

图3-13 J1轴原点设置（设置完成）

1) Press the [4] key on the menu screen, and display the origin/brake selection screen.

在"菜单"界面，按下键盘区的[4]键，选择"原点/制动器"界面（图3-14）。

2) With both hands, slowly move the J2 axis in "+"(plus) direction, and contact the axis against the mechanical stopper.

用双手将J2轴缓慢地向"+"（正）方向移动，直至碰到"机械限位器"。

其余3)~9)步操作与"J1轴"相同。

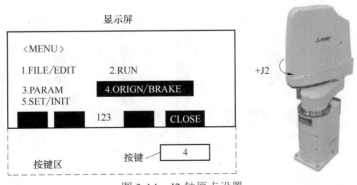

图3-14 J2轴原点设置

(3) J3 and J4 axis origin setting (mechanical stopper)　J3 及 J4 轴原点设置

Always perform origin setting of the J3 axis and the J4 axis simultaneously.

J3 轴与 J4 轴的原点设置必须同时进行。

1) Removes the No. 2 arm cover U。

卸下 2 号机械臂盖板 U。

2) Press the [4] key on the menu screen, and display the origin/brake selection screen.

在菜单界面，按下[4]键，选择"原点 / 制动器"界面（图 3-15）。

3) Press the [2] key, and display the brake release selection screen.

按下[2]键，选择"制动器操作"界面（图 3-16）。

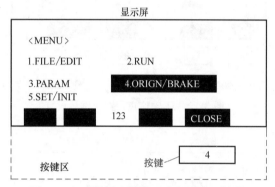
图 3-15　J3 轴原点设置（选择原点 / 制动器）

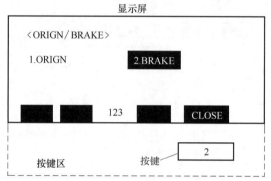
图 3-16　J3 轴原点设置（选择制动器操作）

4) Release the brake of the J3 axis. Input "1" into the J3 axis. Set "0" to other axes.

松开 J3 轴的制动器。按下[↑]~[→]键，将光标移至 J3 的"（ ）"内，按下[1]键。在其他的轴中设置[0]。如图 3-17。

5) Confirm the axis for which the brakes are to be released.

确认要松开制动器的轴。

6) Pressing the [F1] key is kept with the enabling switch of T/B pressed down. The brake is released while pressing the key.

在按下 T/B 的有效开关的状况下持续按压[F1]键。在按压该键期间，制动器被解除（图 3-18）。

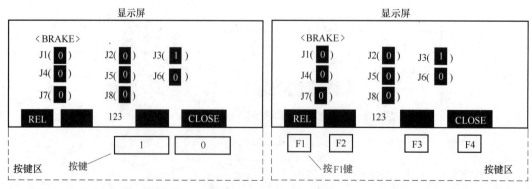

图 3-17　J3 轴原点设置（设置 J3 轴 =1）　　图 3-18　J3 轴原点设置（打开 J3 轴制动器）

Chapter 3　Setting the Robot

Note: The brake of the axis shown below repeats release/lock at the interval in each about 200ms for dropping the J3 axis slowly.

注：为了防止 J3 轴的急剧落下，以约每 200ms 的间隔反复执行制动器解除 / 锁定（间断的制动器解除）。

7) With both hands, slowly move the J3 axis in "+" (plus) direction, and contact the axis against the mechanical stopper.

用双手将 J3 轴缓慢地向 "+"（正）方向移动，碰至机械限位器（图 3-19）。

8) Hold the J4 axis with your hand and rotate it slowly to match the alignment marks. Move the J4 axis with maintaining the condition that the releasing brake of the J3 axis and the J3 axis contact to the mechanical stopper.

在解除 J3 轴制动器的状态下，用手握住 J4 轴，将 J3 轴抵住机械限位器并缓慢转动，对准校准标记或切割线（图 3-20）。

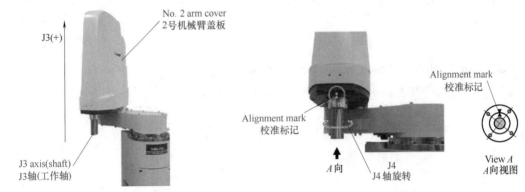

图 3-19　J3 轴原点设置（移动 J3 轴至挡块）　　图 3-20　J4 轴原点设置

9) Detach the [F1] key and work the brake. press the [F4] key and return to the origin/brake screen.

固定完成之后松开 [F1] 键，保持制动。按下 [F4] 键，返回至 "原点 / 制动器" 界面（图 3-21）。

其余操作步骤与 J1 轴相同。

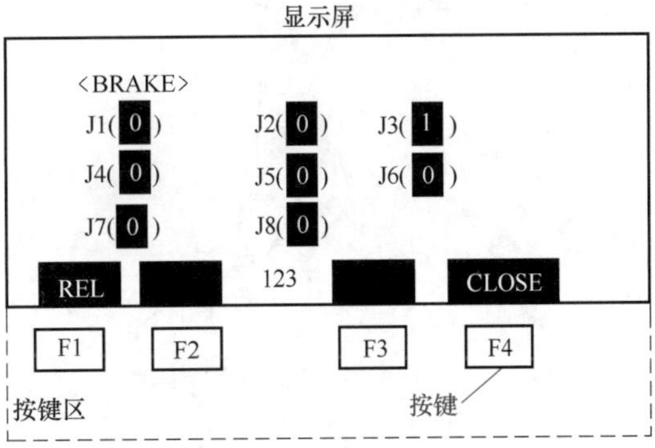

3.2.2　Jig method　校正棒方式

图 3-21　J3 轴制动器有效

This method is using the origin setting tool.

这是使用工具（校正棒）进行原点设置的方式（图 3-22）。

图 3-22　设置原点用校正棒

Setting for 6-axis robot.

6 轴机器人的设置介绍如下。

（1）J1 axis pinhole position　　J1 轴校准孔位置（图 3-23）

（2）J2 axis pinhole position　　J2 轴校准孔位置（图 3-24）

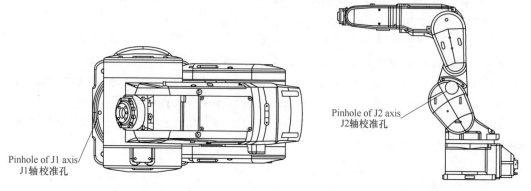

图 3-23　J1 轴校准孔位置　　　　　　图 3-24　J2 轴校准孔位置

（3）J3 axis pinhole position　　J3 轴校准孔位置（图 3-25）

（4）J4 axis pinhole position　　J4 轴校准孔位置（图 3-26）

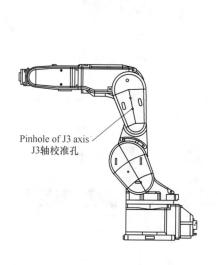 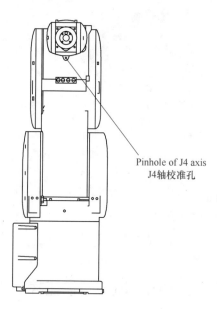

图 3-25　J3 轴校准孔位置　　　　　　图 3-26　J4 轴校准孔位置

（5）J5 axis pinhole position　　J5 轴校准孔位置（图 3-27）

（6）Setting process for J6 axis　　J6 轴设置方法（图 3-28）

Install the bolts (M6×2, customer preparation) in the diagonal position at the J6 axis. Hold the bolts with hands, rotate them slowly and align the ABS mark of the J6 axis with the ABS mark of the wrist area.

如图 3-28，在 J6 轴图示的对角位置安装 2 个螺栓（M6），用手握持螺栓，缓慢旋转，使 J6 轴的 ABS 标志与腕部 ABS 标志对准。

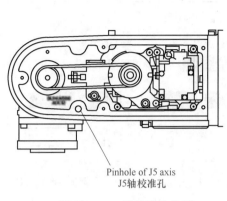
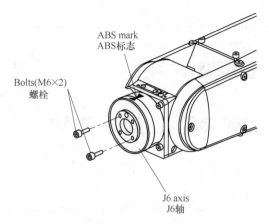

图 3-27　J5 轴校准孔位置　　　　　图 3-28　J6 轴重新设置原点

3.2.3　ABS origin method　ABS 设置原点方式

3.2.3.1　Origin setting for 4-axis robot　4 轴机器人原点设置

（1）J1 axis origin setting　J1 轴原点设置

1）Press the[4]key on the menu screen, and display the origin/brake selection screen.

在菜单界面，按下［4］键，选择"原点／制动器"界面（图 3-29）。

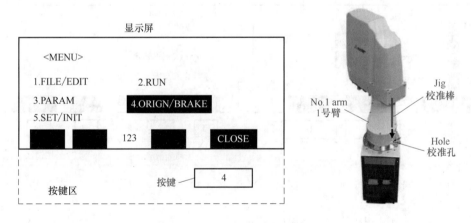

图 3-29　J1 轴原点设置

2）Move the J1 axis slowly toward the front using both hands. Align the pinhole of the No.1 arm and the pinhole at the base section, feed through the origin jig into the pinholes and fasten.

用双手将 J1 轴缓慢移动至前面方向，将基座部的校准孔与 1 号机械臂的校准孔对准后，将校正棒穿过两孔进行固定。

3）Press the[1]key, and display the origin setting selection screen.

按下［1］键，选择"原点设置"界面。如图 3-30。

4）Press the[3]key, and display the tool selection screen.

按下［3］键，选择"校正棒"方式（图 3-31）。

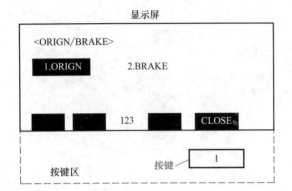 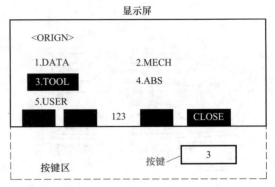

图 3-30　J1 轴原点设置（校正棒方式选择原点设置）　　图 3-31　J1 轴原点设置（校正棒方式选择 TOOL）

5）Input "1" into the J1 axis. Set "0" to other axes.

按下[↑]~[→]键，将光标移至 J1 的"（　）"内，按下[1]键。在其他的轴中设置[0]（图 3-32）。

6）Press the [EXE] key, and display confirmation screen.

按下[EXE]键，显示确认画面。

7）Press the [F1] key, and the origin position is set up.

按下[F1]键，确认设置原点数据（图 3-33）。

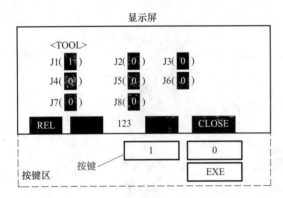 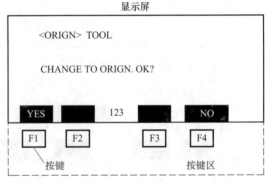

图 3-32　J1 轴原点设置（校正棒方式设置 J1 轴）　　图 3-33　J1 轴原点设置（校正棒方式确认设置）

8）Setting of the origin is completed.

原点设置完成。

（2）J2 axis origin setting　J2 轴原点设置

1）Press the [4] key on the menu screen, and display the origin/brake selection screen.

在菜单界面，按下[4]键，选择"原点/制动器"界面（图 3-34）。

2）Slowly rotate the J2 axis to 0 degree with both hands. And align the pinholes of the No.1 and No.2 arms, feed through the origin jig into the pinholes and fasten.

用双手将 J2 轴缓慢移动至 0°方向。将 2 号机械臂的定位孔与 1 号机械臂的定位孔对准，将校正棒穿过两孔进行固定（图 3-34）。

步骤 3）~8）与 J1 轴相同。

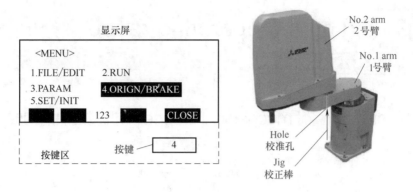

图 3-34　J2 轴原点设置

3.2.3.2　ABS mark attachment positions for 4-axis robot　4 轴机器人 ABS 标志位置（图 3-35）

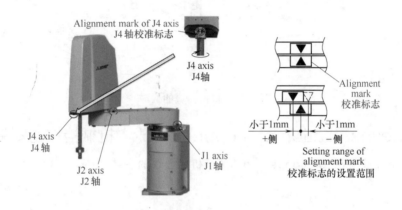

图 3-35　ABS 标志位置

3.2.3.3　ABS mark attachment positions for 6-axis robot　6 轴机器人 ABS 标志位置（图 3-36）

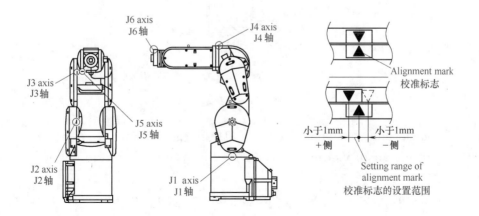

图 3-36　ABS 标志

3.3 Changing the operating range 调整运行范围

3.3.1 The structure of mechanical stopper 机械挡块的构成（图3-37）

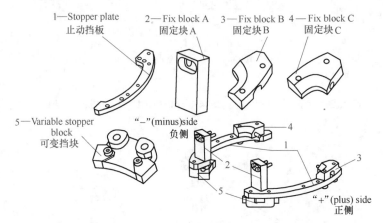

图3-37 机械挡块的构成

3.3.2 The installation of stopper block 机械挡块的安装

（1）The installation of stopper block A 机械挡块的安装A（图3-38）

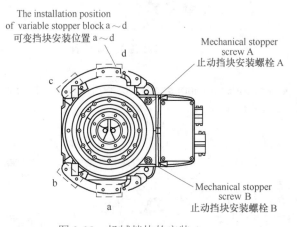

图3-38 机械挡块的安装A

（2）The installation of stopper block B 机械挡块的安装B

The installation procedure of the J1 axis operating range change is shown below. J1轴行程范围调整方法如下。

Fix the fixing block A and the fixing block B to the robot arm as temporary. Fix the fixing block A by using two screws, and fix the fixing block B by using a screw.

Chapter 3　Setting the Robot

将固定块 A 和固定块 B 临时固定在机器人本体上。安装固定块 A 用 2 个螺栓，安装固定块 B 用 1 个螺栓。

1）The installation of stopper block 1. 机械挡块的安装方法 1（图 3-39）。

2）The installation of stopper block 2. 机械挡块的安装方法 2（图 3-40）。

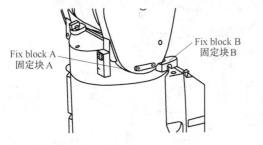 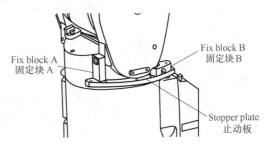

图 3-39　机械挡块的安装方法 1　　　图 3-40　机械挡块的安装方法 2

3）The installation of stopper block 3. 机械挡块的安装方法 3（图 3-41）。

4）The installation of stopper block 4. 机械挡块的安装方法 4（图 3-42）。

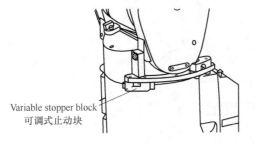 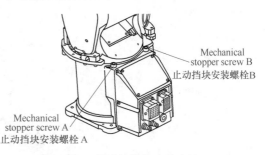

图 3-41　机械挡块的安装方法 3　　　图 3-42　机械挡块的安装方法 4

3.3.3　Installation of J1 axis operating range change option　J1 轴行程挡块的安装

（1）Installation　安装（图 3-43）

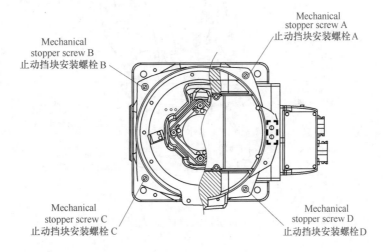

图 3-43　J1 轴行程挡块的安装（13kg）

（2）Steps for J1 axis to adjust travel range　J1 轴调整行程范围步骤

1）Step 1. 步骤 1（图 3-44）。

2）Step 2. 步骤 2（图 3-45）。

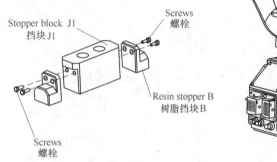

图 3-44　J1 轴机器人调整行程范围步骤 1　　图 3-45　J1 轴机器人调整行程范围步骤 2

3）Step 3. 步骤 3（图 3-46）。

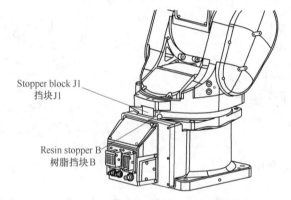

图 3-46　J1 轴机器人调整行程范围步骤 3

4）Step4. 步骤 4（图 3-47）。

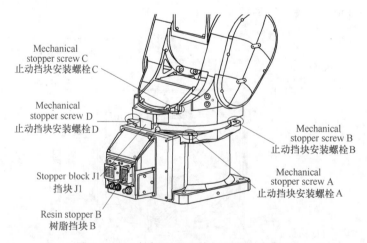

图 3-47　J1 轴机器人调整行程范围步骤 4

Chapter 4
How to Operate Robot
机器人的操作

4.1 Operation panel (O/P) functions
操作面板的使用

Name of each part. 控制面板各部分名称（图 4-1）。

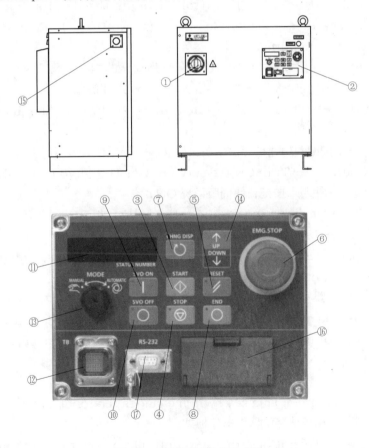

图 4-1　控制面板各部分名称

① Power switch——Turn the control power ON/OFF (with earth leakage breaker function).

控制器的电源开关（带漏电保护功能）。

② Operation panel——The operation panel for servo ON/OFF, START/STOP the program, etc.

操作面板——用于执行伺服 ON/OFF，启动/停止程序等操作。

③ START button——Execute the program and operates the robot. The program is run continuously.

启动按键——用于启动程序，程序是连续执行的。

④ STOP button——Stop the robot immediately. The servo do not turn OFF.

停止按键——立即停止机器人运行。但伺服系统不处于 OFF 状态。

⑤ RESET button——Reset the error. Also reset the program's halted state and resets the program.

复位按键——解除故障报警，也解除程序的暂停状态并使程序复位。

⑥ Emergency stop switch——Switch stop the robot in an emergency state. The servo turns OFF.

急停开关——使用急停开关使机器人处于紧急停止。伺服系统 OFF。

⑦ CHNGDISP button——This button changes the details displayed on the display panel in the order of "Override" → "Line No." → "Program No." → "User information" → "Maker information".

显示转换按键——使用本开关依次改换显示屏所显示的内容，显示内容依次为"速度倍率"→"程序行号"→"程序号"→"用户信息"→"制造商信息"。

⑧ END button——Stop the program being executed at the last line or END statement.

END 按键——本按键的功能是使程序在执行最后一行或在 END 行停止。

⑨ SVO.ON button——Turn ON the servo power.The servo turns ON.

伺服 ON 按键——使用本按键，使伺服系统 ON。

⑩ SVO.OFF button——Turn OFF the servo power.The servo turns OFF.

伺服 OFF 按键——使用本按键，使伺服系统 OFF。

⑪ Display panel (STATUS.NUMBER) ——The alarm number, program number, override value (%), etc, are displayed.

显示屏——显示报警号、程序号、速度倍率等内容。

⑫ T/B connection connector (TB) ——This is a dedicated connector for connecting the T/B. When not using T/B, connect the attached dummy connector.

T/B 示教单元插口——本插口规定用于连接示教单元。不使用示教单元时，必须安装一空插头。

⑬ Mode key switch——This key switch changes the robot's operation mode.

模式选择开关——用于选择机器人操作模式。

AUTOMATIC——Operations from the controller or external equipment are valid. Operations for which the operation mode must be at the external device or T/B are not possible (exclude the start of automatic operation).

自动模式——对于从控制器或外部设备发出指令的操作是有效的。这种操作模式必须使用外部操作信号，使用T/B操作单元无效（除自动启动操作）。

MANUAL——When the T/B is valid, only operations from the T/B are valid. Operations for which the operation mode must be at the external device or controller are not possible.

手动操作——当手持单元T/B有效时，仅仅只有T/B发出的指令有效，而其他外部器件或控制器的操作无效。

⑭ UP/DOWN button——This button scrolls up or down the details displayed on the "STATUS.NUMBER" display panel.

上/下滚动翻页键——使用本按键，在显示屏上/下翻页显示"状态/数字"内容。

⑮ Cable lead-in port——Draw in the primary power cable.

电缆口——电源电缆进出口。

⑯ Interface cover——USB interface and battery are mounted.

盖板——内部有USB插口和电池。

⑰ RS-232 connector——This is an RS-232 specification connector for connecting the personal computer.

RS-232插口——连接计算机的RS-232口。

4.2 Installation of teaching pendant 示教单元的安装

（1）Installation of teaching pendant 1　安装连接示教单元1（图4-2）

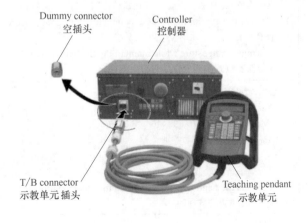

图4-2　安装连接示教单元

（2）Installation of teaching pendant 2　安装连接示教单元2（图4-3）

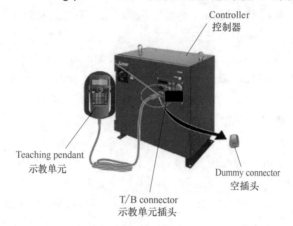

图4-3　示教单元的安装

4.3　Explanation of operation methods 操作方式说明（表4-1）

表4-1　点动模式类型及应用

Jog mode 点动模式	Main application 主要应用
JOINT jog 关节型点动	• Moves each joint　各轴以关节旋转方式运动 • Moves the robot arm largely　使机器人在最大范围运动 • Changes the robot posture　改变机器人的"立体形位"
XYZ jog 直交型点动	• Accurately sets the teaching position 精确设置示教位置 • Moves the axis straight along the *XYZ* coordinate system 或在*XYZ*坐标系内以直线方式移动各轴 • Moves the axis straight while maintaining the robot posture 保持机器人"立体形位"以直线方式移动各轴 • Changes the posture while maintaining the hand position 保持抓手位置改变"立体形位"
TOOL jog 以工具坐标系 为基准点动	• Accurately sets the teaching position 精确设置示教位置 • Moves the axis straight along the hand direction 沿抓手方向直线移动各轴 • Changes the posture while maintaining the hand position 保持抓手位置改变"立体形位" • Rotates the hand while maintaining the hand position 保持抓手位置旋转抓手
3-axis XYZ jog 三轴型点动	• When the axis cannot be moved with *XYZ* jog that maintains the posture 当不能够以*XYZ*点动方式移动保持"立体形位"时 • When the tip is to be moved linearly but the posture is to be changed. 当抓手前端需要保持直线移动同时需要改变"立体形位"时

续表

Jog mode 点动模式	Main application 主要应用
CYLINDER jog 圆柱型点动	• Moves in a cylindrical shape centering on the Z axis while maintaining the posture. 需要保持"立体形位"而在一个以 Z 轴为中心的圆柱形上运动时 • Moves linearly in a radial shape centering on the Z axis while maintaining the posture 需要保持"立体形位"而在一个以 Z 轴为中心的径向形状上直线运动时
WORK jog （work jog mode） 在工件坐标系中的点动	• Accurately sets the teaching position 精确设置示教位置 • Moves the axis straight along the coordinate system（work coordinate system）defined in accordance with a workpiece, pallet, etc 需要在一个由工件、托盘等定义的坐标系（工件坐标系）内直线运动时 • Changes the posture along the work coordinate system 沿工件坐标系改变"立体形位"时
WORK jog （Ex-T jog mode） 在 Ex-T 坐标系 中的点动	• Accurately sets the teaching position 精确设置示教位置 • Moves the axis straight along the work coordinate system（Ex-T coordinate system）defined in accordance with an installed grinder, dispenser, etc 需要在一个工件坐标系（Ex-T 坐标系）内直线运动时。Ex-T 坐标系与外部安装的研磨机、点胶机等相对应 • Changes the posture along the work coordinate system（Ex-T coordinate system） 沿工件坐标系（Ex-T 坐标系）改变"立体形位"时

4.3.1　JOINT jog　关节型点动

In this mode, each axis moves independently; each of the axes can be adjusted independently. It is possible to adjust the coordinates of the axes J1 to J6 as well as the additional axes J7 and J8 independently.

在这种模式中，各轴独立运行，每一轴都能独立调节。可以独立地调节 J1 ~ J6 轴以及附加轴 J7 ~ J8 轴。

（1）JOINT jog for horizontal type robot　水平型机器人关节型点动（图 4-4）

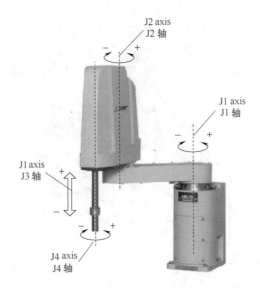

图 4-4　关节型点动

1）J1 axis jog operation　J1 轴点动（图 4-5）

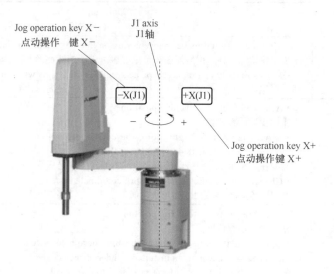

图 4-5　J1 轴关节型点动

When the [+X（J1）] key is pressed, the J1 axis will rotate in the plus direction. When the [-X（J1）] key is pressed, rotate in the minus direction.

按下 [+X（J1）] 键，J1 轴正向旋转。按下 [-X（J1）] 键，J1 轴负向旋转。

2）J2 axis jog operation　J2 轴点动（图 4-6)

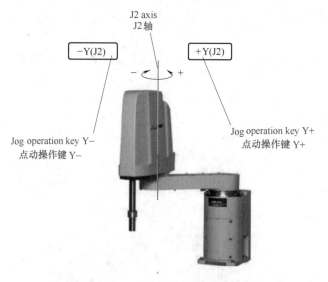

图 4-6　J2 轴关节型点动

When the [+Y（J2）] key is pressed, the J2 axis will rotate in the plus direction. When the [-Y（J2）] key is pressed, rotate in the minus direction.

按下 [+Y（J2）] 键，J2 轴正向旋转。按下 [-Y（J2）] 键，J2 轴负向旋转。

3) J3 axis jog operation　J3 轴点动（图 4-7）

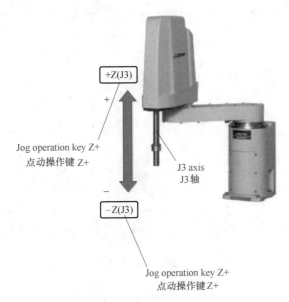

图 4-7　J3 轴关节型点动

When the [+Z（J3）] key is pressed, the J3 axis will rotate in the plus direction. When the [-Z（J3）] key is pressed, rotate in the minus direction.

按下 [+Z（J3）] 键，J3 轴正向运动。按下 [-Z（J3）] 键，J3 轴负向运动。

4) J4 axis jog operation　J4 轴点动（图 4-8）

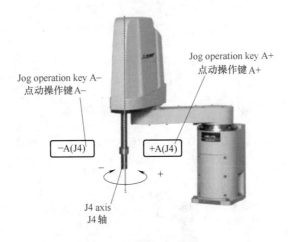

图 4-8　J4 轴关节型点动

When the [+A（J4）] key is pressed, the J4 axis will rotate in the plus direction. When the [-A（J4）] key is pressed, rotate in the minus direction.

按下 [+A（J4）] 键，J4 轴正向旋转运动。按下 [-A（J4）] 键，J4 轴负向旋转运动。

(2) Vertical type robot JOINT jog　垂直型机器人关节型点动

Adjusts the coordinates of each axis independently in angle units.

各轴分别以角度单位运动。

1）Vertical type robot JOINT jog　垂直型机器人关节型点动（图4-9）

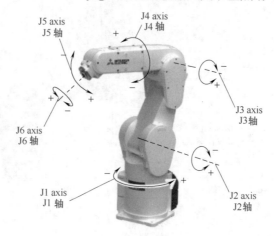

图4-9　垂直型机器人关节型点动

2）JOINT jog for J1 axis of vertical type robot　垂直型机器人J1轴关节型点动（图4-10）

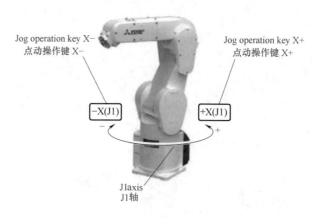

图4-10　垂直型机器人J1轴关节型点动

3）JOINT jog for J2 axis of vertical type robot　垂直型机器人J2轴关节型点动（图4-11）

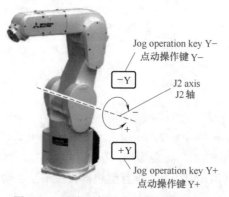

图4-11　垂直型机器人J2轴关节型点动

4）JOINT jog for J3 axis of vertical type robot　垂直型机器人J3轴关节型点动（图4-12）

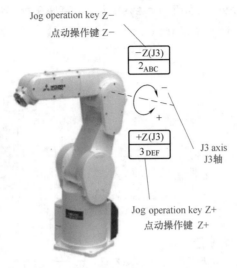

图4-12　垂直型机器人J3轴关节型点动

5）JOINT jog for J4, J5 and J6 axis of vertical type robot　垂直型机器人J4、J5、J6轴关节型点动（图4-13）

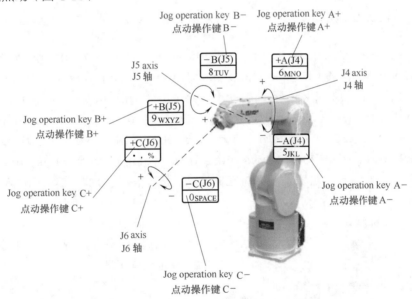

图4-13　垂直型机器人J4、J5、J6轴关节型点动

4.3.2　XYZ jog　直交型点动

Adjusts the axis coordinates along the direction of the robot coordinate system. The X, Y, and Z axis coordinates are adjusted in "mm" units. The A, B, and C axis coordinates are adjusted in angle units.

沿着机器人坐标系的方向移动各轴。X、Y、Z轴以mm为单位移动。A、B、C轴以角度为单位移动。

4.3.2.1 XYZ jog for horizontal type robot 水平型机器人 XYZ 型点动

（1）XYZ jog for horizontal type robot 水平型机器人 XYZ 型点动（图 4-14）

（2）Moving along the base coordinate system 沿基座坐标系移动（图 4-15）

The direction of the end axis will not change.

工作轴方位不变。

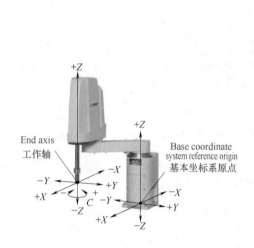
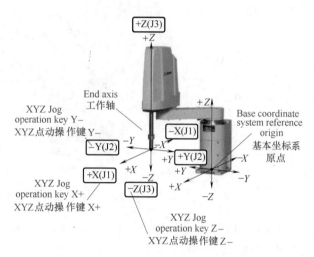

图 4-14 XYZ 型点动 1　　　　图 4-15 XYZ 型点动 2（沿基座坐标系移动）

When the [+X（J1）] key is pressed, the robot will move along the X axis plus direction. When the [-X（J1）] key is pressed, move along the minus direction.

When the [+Y（J2）] key is pressed, the robot will move along the Y axis plus direction. When the [-Y（J2）] key is pressed, move along the minus direction.

When the [+Z（J3）] key is pressed, the robot will move along the Z axis plus direction. When the [-Z（J3）] key is pressed, move along the minus direction.

按下 [+X（J1）] 键，机器人沿 X 轴正向运动。按下 [-X（J1）] 键，机器人沿 X 轴负向运动。

按下 [+Y（J2）] 键，机器人沿 Y 轴正向运动。按下 [-Y（J2）] 键，机器人沿 Y 轴负向运动。

按下 [+Z（J3）] 键，机器人沿 Z 轴正向运动。按下 [-Z（J3）] 键，机器人沿 Z 轴负向运动。

（3）Changing the end axis posture 改变工作轴的形位（发生了旋转）

The end axis position point is not changing.

工作轴的"位置点"不变（图 4-16）。

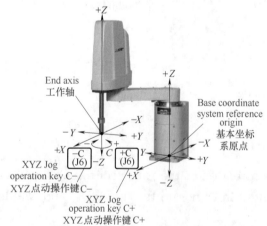

图 4-16 XYZ 型点动 3

4.3.2.2　XYZ jog for vertical type robot　垂直型机器人 XYZ 型点动

（1）XYZ jog for vertical type robot　垂直型机器人 XYZ 型点动（图 4-17）

（2）Select XYZ jog　选择 jog 模式

Select XYZ jog with TB.

1) Press "JOG" key.

2) Press "F1" key.

The screen display is shown in Fig.4-18.

在手持操作单元上选择 XYZ jog 模式。

1）按"JOG"键；

2）按"F1"键。

画面显示见图 4-18。

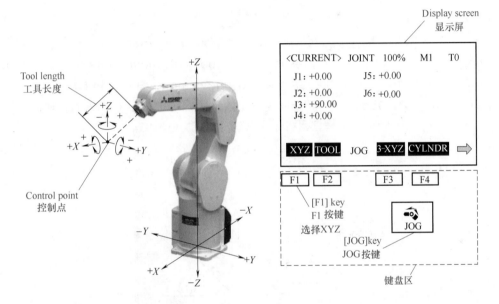

图 4-17　垂直型机器人 XYZ 型点动　　图 4-18　手持操作单元的显示与操作

（3）XYZ jog operation 1　垂直型机器人 XYZ 型点动（法兰方向不变）

Moving along the base coordinate system, the direction of the flange will not change.

按基本坐标系运动，法兰方向不变（图 4-19）。

Operating with T/B　使用手持单元进行操作：

When the [+X（J1）] key is pressed, the robot will move along the X axis plus direction. When the [-X（J1）] key is pressed, move along the minus direction.

按 [+X（J1）] 键，机器人向 X 轴正向运动。按 [-X（J1）] 键，机器人向 X 轴负向运动。

When the [+Y（J2）] key is pressed, the robot will move along the Y axis plus direction. When the [-Y（J2）] key is pressed, move along the minus direction.

按 [+Y（J2）] 键，机器人向 Y 轴正向运动。按 [-Y（J2）] 键，机器人向 Y 轴负向运动。

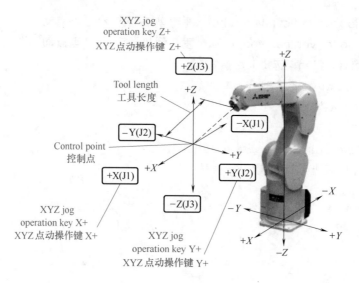

图 4-19 垂直型机器人 XYZ 型点动（沿基本坐标系）

When the [+Z（J3）] key is pressed, the robot will move along the Z axis plus direction. When the [-Z（J3）] key is pressed, move along the minus direction.

按 [+Z（J3）] 键，机器人向 Z 轴正向运动。按 [-Z（J3）] 键，机器人向 Z 轴负向运动。

（4）XYZ jog operation 2　垂直型机器人 XYZ 型点动（图 4-20）

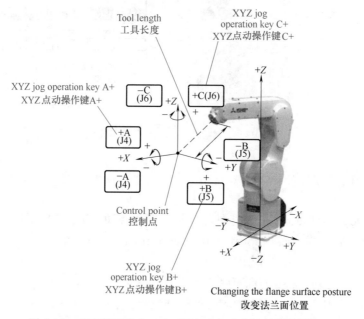

图 4-20 垂直型机器人 XYZ 型点动（改变法兰面位置）

When the [+A（J4）] key is pressed, the X axis will rotate in the plus direction. When the [-A（J4）] key is pressed, rotate in the minus direction.

按 [+A（J4）] 键，绕 X 轴正向旋转。按 [-A（J4）] 键，绕 X 轴负向旋转。

When the [+B（J5）] key is pressed, the Y axis will rotate in the plus direction. When the [-B（J5）] key is pressed, rotate in the minus direction.

按 [+B（J5）] 键，绕 Y 轴正向旋转。按 [-B（J5）] 键，绕 Y 轴负向旋转。

When the [+C (J6)] key is pressed, the Z axis will rotate in the plus direction. When the [-C (J6)] key is pressed, rotate in the minus direction.

按 [+C（J6）] 键，绕 Z 轴正向旋转。按 [-C（J6）] 键，绕 Z 轴负向旋转。

4.3.3　TOOL jog　以工具坐标系为基准点动

The position can be adjusted forward/backward, left/right, or to the upward/downward relative to the direction of the hand tip of the robot (the tool coordinate system).

TOOL jog 模式可以相对于机器人抓手前端（即工具坐标系）做前后、左右、上下的位置调节。

The tip moves linearly. The posture can be rotated around the X, Y, and Z axes of the tool coordinate system of the hand tip by pressing the A, B, and C keys, without changing the actual position of the hand tip. It is necessary to specify the tool length in advance using the MEXTL parameter.

抓手前端可做直线运动。改变抓手的"立体形位"也可以通过旋转工具坐标系的 $X/Y/Z$ 轴来实现，由操作 $A/B/C$ 键执行旋转，而且可以不改变抓手工作点的实际工作位置。在进一步的使用中需要使用参数 MEXTL 设置抓手的长度。

The tool coordinate system, in which the hand tip position is defined, depends on the type of robot. In the case of a vertical multi-joint type robot, the direction from the mechanical inter face plane to the hand tip is +Z. In the case of a horizontal multi-joint type robot, the upward direction from the mechanical interface plane is +Z.

在工具坐标系中，抓手工作点的位置被确定。对于垂直型多轴机器人，从机械接口平面对着抓手工作点的方向是 Z+。对于水平型多轴机器人，机械接口平面朝上的方向是 Z+。

4.3.3.1　TOOL jog for horizontal type robot　水平型机器人 TOOL 型点动

（1）TOOL jog for horizontal type robot　水平型机器人 TOOL 型点动（图 4-21）

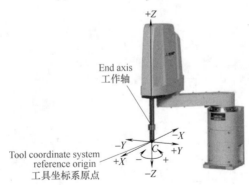

图 4-21　TOOL 型点动

（2）Select of method of operation　选择操作方法

Select TOOL jog with TB

1）Press "JOG" key;

2）Press "F2" key.

The screen display is shown in Fig. 4-22.

在手持操作单元上选择 TOOL jog 模式。

1）按"JOG"键；

2）按"F2"键。

画面显示见图 4-22。

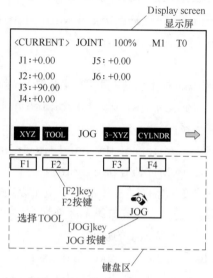

图 4-22　选择 4 轴机器人 TOOL 型点动

（3）Moving along the tool coordinate system, the direction of the end axis will not change　在工具坐标系中运动，工作轴方位不变（图 4-23）

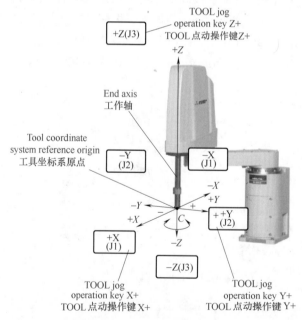

图 4-23　TOOL 型点动工作轴方向不变

（4）Changing the end axis posture, the Position of the end axis will not change　改变工作轴的方位，但工作轴的位置不变（图 4-24）

When the[+C（J6）] key is pressed, the Z axis will rotate in the plus direction of the tool coordinate system.When the[-C (J6)] key is pressed, rotate in the minus direction.

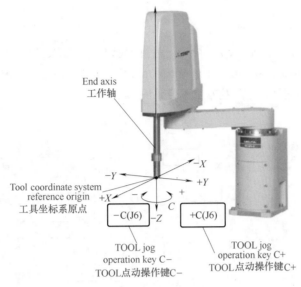

图 4-24 TOOL 型点动

按下 [+C (J6)] 键，Z 轴按工具坐标系的正向旋转。按下 [-C (J6)] 键，Z 轴按工具坐标系的负向旋转。

4.3.3.2 TOOL jog for vertical type robot　垂直型机器人 TOOL jog

(1) Select TOOL jog　选择 TOOL 型点动

Select TOOL jog with TB.

1) Press "JOG" key.
2) Press "F2" key.

The screen display is shown in Fig. 4-25.

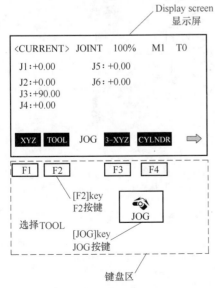

图 4-25 选择 TOOL 型点动

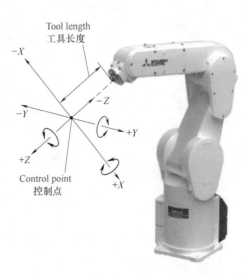

图 4-26 垂直型机器人 TOOL 型点动

在手持操作单元上选择 TOOL jog 模式。

1）按"JOG"键；

2）按"F2"键。

画面显示见图 4-25。

（2）TOOL jog for vertical type robot　垂直型机器人 TOOL 型点动（图 4-26）

（3）Moving along the tool coordinate system,the direction of the flange will not change 在工具坐标系内运动，法兰方位不变（图 4-27）

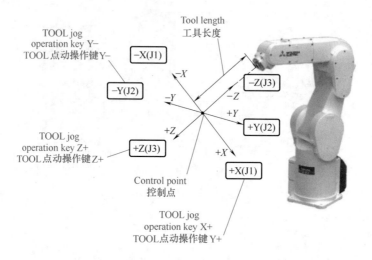

图 4-27　垂直型机器人 TOOL 型点动（沿工具坐标系法兰方向不变）

（4）Changing the flange surface posture，the control point does not change　改变法兰面方位，控制点不变（图 4-28）

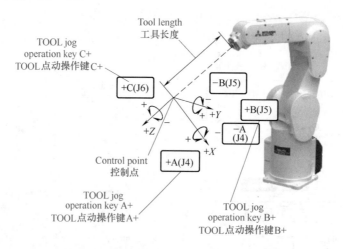

图 4-28　垂直型机器人 TOOL 型点动（改变法兰面方位）

4.3.4　3-axis XYZ jog　三轴型点动

（1）Outline　概述

The axes are adjusted linearly with respect to the robot coordinate system.

Chapter 4 How to Operate Robot

在 3-axis XYZ 点动模式中，各轴线性运动以机器人坐标系为基准。

Unlike in the case of XYZ jog, the posture will be the same as in the case of the J4, J5, and J6 axes JOINT jog feed.

与 XYZ 点动模式不同，改变"立体形位"的操作与关节型点动的 J4/J5/J6 轴操作一致。

While the position of the hand tip remains fixed, the posture is interpolated by X, Y, Z, J4, J5, and J6; i.e., a constant posture is not intained. It is necessary to specify the tool length in advance using the MEXTL parameter.

当抓手工作点的位置保持固定时，由 X, Y, Z, J4, J5, 和 J6 规定了对"立体形位"的操作，即不可能保持连续的"立体形位"。在进一步的使用中需要使用参数 MEXTL 设置抓手的长度。

In the direction of $X/Y/Z$, the 3-axis JOG is based on the "world coordinate system". The unit of movement is mm. But the movement of the $A/B/C$ axis corresponds to the J4/J5/J6 axis. The unit of movement is angle. The 3-axis JOG combines the advantages of the two coordinate systems.

三轴直交型 JOG 在 $X/Y/Z$ 方向上是以"世界坐标系"为基准，移动单位是 mm。但是 $A/B/C$ 三轴的移动则是对应 J4/J5/J6 轴，以角度为单位。这种方式综合了两种坐标系的优势。

（2）Select 3-axis XYZ jogt　选择 3-axis XYZ jog 点动

Select 3-axis XYZ jog with TB.

1）Press "JOG" key；

2）Press "F3" key.

The screen display is shown in Fig. 4-29.

在手持操作单元上选择 3-axis XYZ jog 模式。

1）按"JOG"键；

2）按"F3"键。

画面显示见图 4-29。

4.3.4.1 3-axis XYZ jog for horizontal type robot　水平型机器人三轴型点动

Moving along the base coordinate system, the direction of the end axis will change.

按基本坐标系运动，改变工作轴方位（图 4-30）。

When the [+X（J1）] key is pressed, the robot will move along the X axis plus direction. When the [-X（J1）] key is pressed, move along the minus direction.

When the [+Y（J2）] key is pressed, the robot will move along the Y axis plus direction. When the [-Y（J2）] key is pressed, move along the minus direction.

When the [+Z（J3）] key is pressed, the robot will move along the Z axis plus direction. When the [-Z（J3）] key is pressed, move along the minus direction.

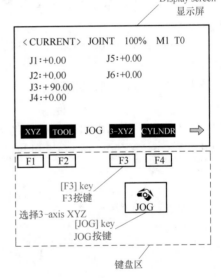

图 4-29 手持单元选择三轴 XYZ 型点动

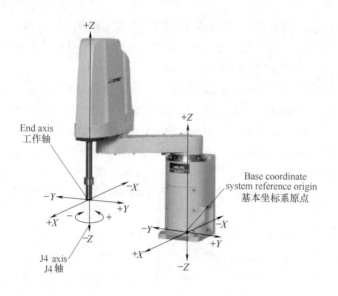

图 4-30　三轴型点动（工作轴改变方位）

按下 [+X（J1）] 键，机器人沿 X 轴正向运动。按下 [-X（J1）] 键，机器人沿 X 轴负向运动。

按下 [+Y（J2）] 键，机器人沿 Y 轴正向运动。按下 [-Y（J2）] 键，机器人沿 Y 轴负向运动。

按下 [+Z（J3）] 键，机器人沿 Z 轴正向运动。按下 [-Z（J3）] 键，机器人沿 Z 轴负向运动。

4.3.4.2　3-axis XYZ jog for vertical type robot　垂直型机器人 3-XYZ 点动

（1）3-axis XYZ jog for vertical type robot　垂直型机器人 3-XYZ 点动（图 4-31）

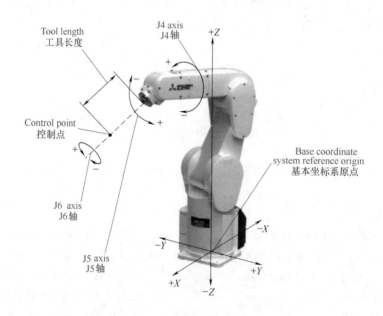

图 4-31　垂直型机器人 3-XYZ 点动

(2) Moving along the base coordinate system, the direction of the flange will not change 在基本坐标系内运动，法兰方向不变（图4-32）

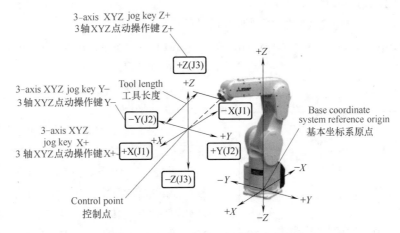

图 4-32　3-XYZ 点动（沿基本坐标系）

(3) Changing the flange surface posture 改变法兰面方位（图4-33）

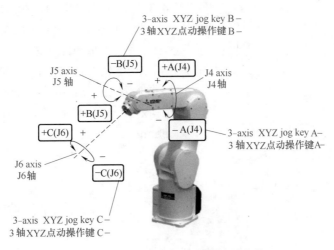

图 4-33　3-XYZ 点动（改变法兰面方位）

4.3.5　CYLINDER jog　圆柱型点动

Use the cylindrical jog when moving the hand in the cylindrical direction with respect to the robot's origin.

当移动抓手在圆柱型空间中移动时，使用"圆柱型点动模式"，"圆柱型空间基准"与机器人原点相同。

Adjusting the X-axis coordinate moves the hand in the radial direction from the center of the robot.

X 轴坐标表示以机器人中心为圆柱中心的径向坐标。

Adjusting the Y-axis coordinate moves the hand in the same way as in JOINT jog feed around the J1 axis.

Y 坐标与关节插补的 J1 轴相同，表示旋转角度。

Adjusting the Z-axis coordinate moves the hand in the Z direction in the same way as in XYZ jog feed.

Z 坐标与 XYZ 点动模式中的 Z 坐标相同。

Adjusting the coordinates of the A, B, and C axes rotates the hand in the same way as in XYZ jog feed.

$A/B/C$ 轴的坐标与 XYZ 点动模式中的 $A/B/C$ 轴坐标相同。

（参考：圆筒型 JOG 首先要建立一个圆筒型坐标系。在圆筒型坐标系中，X 坐标表示了圆筒的半径，Z 坐标表示圆筒的高度，Y 坐标表示了圆筒的旋转角度，也就是 J1 轴的角度。其余 $A/B/C$ 轴的旋转方向与 XYZ 坐标系相同。这样圆筒型 JOG 就相当于机器人控制点在一个圆筒壁上做运动。或者说，如果是一个圆筒壁上的运动，就选取圆筒型 JOG 最为适宜。）

4.3.5.1 CYLINDER jog for horizontal type robot 水平型机器人的圆柱型点动

（1）Selection 选择（图 4-34）

（2）CYLINDER jog 圆柱型点动（图 4-35）

Moving along an arc centering on the Z axis, the direction of the flange will not change.

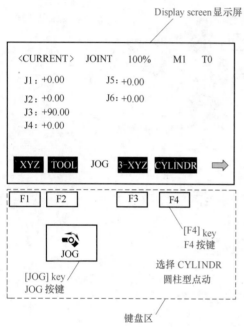

图 4-34 选择圆柱型点动

沿圆柱面运动，法兰方位不变。

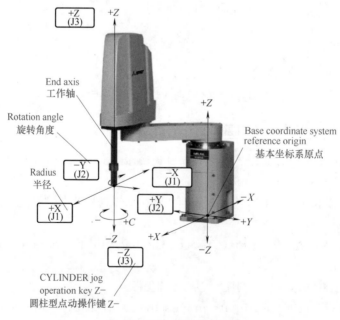

图 4-35 圆柱型点动

When the[+X（J1）] key is pressed, the robot will expand in the radial direction. When the[-X（J1）] key is pressed, contract in the radial direction.

When the[+Y（J2）] key is pressed, the robot will move along the arc in the plus direction. When the[-Y（J2）] key is pressed, move in the minus direction.

When the[+Z（J3）] key is pressed, the robot will move along the Z axis plus direction. When the[-Z（J3）] key is pressed, move along the minus direction.

按下 [+X（J1）] 键，机器人沿半径增加方向移动。按下 [-X（J1）] 键，机器人沿半径减小方向移动。

按下 [+Y（J2）] 键，机器人沿圆弧正向移动。按下 [-Y（J2] 键，机器人沿圆弧负向移动。

按下 [+Z（J3）] 键，机器人沿 Z 轴正向移动。按下 [-Z（J3] 键，机器人沿 Z 轴负向移动。

（3）Changing the flange surface posture, the position of the end axis will not change 法兰方位改变，工作轴（直角坐标）位置不变（图 4-36）

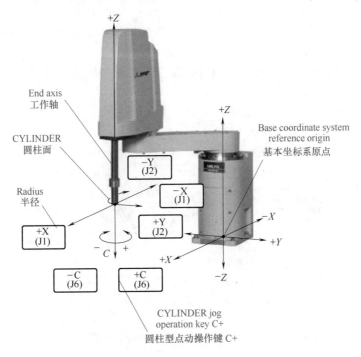

图 4-36　圆柱型点动

When the [+C（J6）] key is pressed, the Z axis will rotate in the plus direction. When the [-C（J6）] key is pressed, rotate in the minus direction.

按下 [+C（J6）] 键，Z 轴正向旋转。按下 [-C（J6）] 键，Z 轴负向旋转。

4.3.5.2　CYLINDER jog for vertical type robot　垂直型机器人圆柱型点动

（1）CYLINDER jog operation　圆柱型点动（图4-37）

（2）Moving along an arc centering on the Z axis　沿圆柱中心为Z轴的圆柱面运动（图4-38）

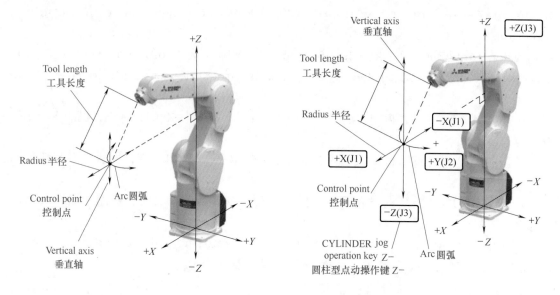

图4-37　垂直型机器人圆柱型点动　　图4-38　垂直型机器人圆柱型点动（沿圆柱中心运行）

（3）Changing the flange surface posture　改变法兰面方位（图4-39）

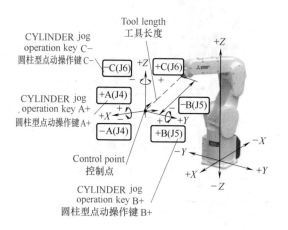

图4-39　垂直型机器人圆柱型点动（改变法兰面方位）

When the [+A（J4）] key is pressed, the X axis will rotate in the plus direction. When the [-A（J4）] key is pressed, rotate in the minus direction.

When the [+B（J5）] key is pressed, the Y axis will rotate in the plus direction. When the [-B（J5）] key is pressed, rotate in the minus direction.

When the [+C（J6）] key is pressed, the Z axis will rotate in the plus direction. When the [-C（J6）] key is pressed, rotate in the minus direction.

按下 [+A（J4）] 键，绕 X 轴正向旋转。按下 [-A（J4）] 键，绕 X 轴负向旋转。
按下 [+B（J5）] 键，绕 Y 轴正向旋转。按下 [-B（J5）] 键，绕 Y 轴负向旋转。
按下 [+C（J6）] 键，绕 Z 轴正向旋转。按下 [-C（J6）] 键，绕 Z 轴负向旋转。

4.3.6　WORK jog　在工件坐标系中的点动

It is necessary to set "0（Work jog mode）" in the parameter WKnJOGMD（n = 1 to 8）in advance to perform this jog operation.

在使用 WORK jog 模式执行点动操作前，必须预先设置参数 WKnJOGMD=0（n 为 1 ~ 8）。

The axes are adjusted linearly with respect to the work coordinate system. The posture rotates around the X, Y, and Z axes of the work coordinate system by pressing the A, B, and C keys, without changing the actual position of the hand tip.

各轴的直线运动以"工件坐标系"为基准。旋转运动为绕"工件坐标系"的 X/Y/Z 轴运动。按 A/B/C 键执行，但不改变抓手工作点的实际位置。

It is necessary to specify the tool length in advance using the MEXTL parameter.

必须使用参数 MEXTL 设置抓手长度。

4.3.6.1　WORK jog for horizontal type robot　水平型机器人在工件坐标系中的点动

（1）WORK jog　在工件坐标系中点动（图 4-40）

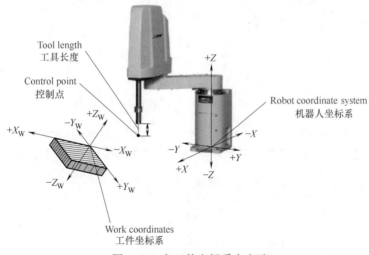

图 4-40　在工件坐标系中点动

When the [+X（J1）] key is pressed, the robot will move along the X axis（X_W）plus direction on the work coordinate system. When the [-X（J1）] key is pressed, move along the minus direction.

When the [+Y（J2）] key is pressed, the robot will move along the Y axis（Y_W）plus direction on the work coordinate system. When the [-Y（J2）] key is pressed, move along the minus direction.

When the [+Z（J3）] key is pressed, the robot will move along the Z axis（Z_W）plus direction on the work coordinate system. When the [-Z（J3）] key is pressed, move along the minus direction.

按下 [+X（J1）] 键，机器人沿工件坐标系的 X 轴正向运动。按下 [-X（J1）] 键，机器人沿工件坐标系的 X 轴负向运动。

按下 [+Y（J2）] 键，机器人沿工件坐标系的 Y 轴正向运动。按下 [-Y（J2）] 键，机器人沿工件坐标系的 Y 轴负向运动。

按下 [+Z（J3）] 键，机器人沿工件坐标系的 Z 轴正向运动。按下 [-Z（J3）] 键，机器人沿工件坐标系的 Z 轴负向运动。

（2）Changing the end axis posture　改变工作轴方位（Z 轴旋转）

Changing the end axis posture, the position of the control point does not change.

改变工作轴方位（Z 轴旋转），但控制点的位置不变（图 4-41）。

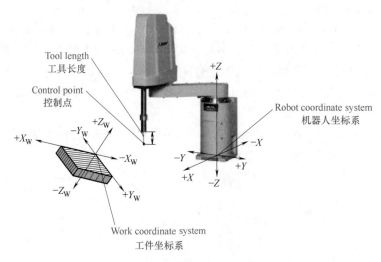

图 4-41　在工件坐标系中点动（改变工作轴方位）

（3）Changing the robot posture in work coordinate system（Ex-T coordinate system）　在工件坐标系（Ex-T 坐标系）中点动改变"立体形位"（图 4-42）

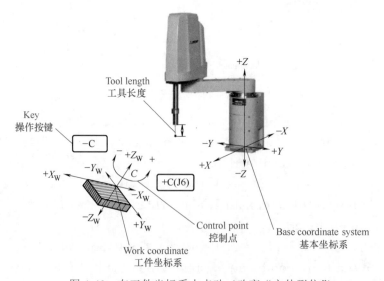

图 4-42　在工件坐标系中点动（改变"立体形位"）

When the [+C (J6)] key is pressed, the control point will rotate in the plus direction around the Z axis (Z_W) of work coordinate system (Ex-T coordinate system). When the [−C (J6)] key is pressed, the control point will rotate in the minus direction.

按下 [+C（J6）] 键，控制点绕工件坐标系（Ex-T 坐标系）Z 轴正向旋转。按下 [−C（J6）] 键，控制点绕工件坐标系（Ex-T 坐标系）Z 轴负向旋转。

4.3.6.2 WORK jog for vertical type robot 垂直型机器人在工件坐标系中的点动

(1) Setting of the work coordinate system 设置工件坐标系（图 4-43）

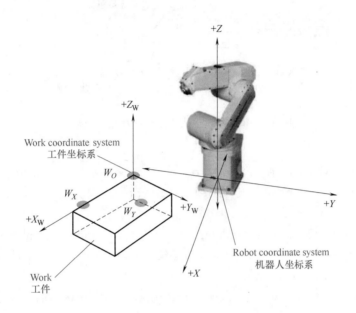

图 4-43 在工件坐标系中点动

Teaching point:

W_O: Work coordinate system origin.

W_X: Position on the "+X" axis of work coordinate system.

W_Y: Position at the side of "+Y" axis on the X-Y plane of work coordinate system.

示教点：

W_O：工件坐标系原点。

W_X：工件坐标系 +X 轴上的位置。

W_Y：工件坐标系的 X-Y 平面上 +Y 轴上的位置。

(2) WORK jog 在工件坐标系中的点动

The jog movement based on work coordinate system, the direction of the flange will not change.

沿工件坐标系运动，法兰方向不变。

The direction of the flange will not change. Move the control point with a straight line in accordance with the work coordinate system.

如图 4-44，法兰方向不变。控制点按工件坐标系做直线运动。

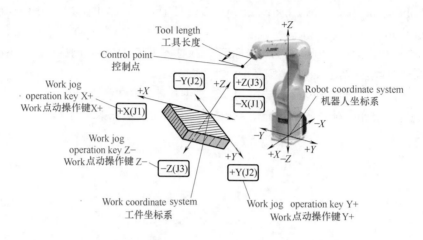

图 4-44　垂直型机器人在工件坐标系的点动

When the [+X（J1）] key is pressed, the robot will move along the X axis (X_W) plus direction on the work coordinate system. When the [-X（J1）] key is pressed, move along the minus direction.

When the [+Y（J2）] key is pressed, the robot will move along the Y axis (Y_W) plus direction on the work coordinate system. When the [-Y（J2）] key is pressed, move along the minus direction.

When the [+Z（J3）] key is pressed, the robot will move along the Z axis (Z_W) plus direction on the work coordinate system. When the [-Z（J3）] key is pressed, move along the minus direction.

按下 [+X（J1）] 键，机器人沿工件坐标系的 X 轴（X_W）正向运动。按下 [-X（J1）] 键，机器人沿工件坐标系的 X 轴（X_W）负向运动。

按下 [+Y（J2）] 键，机器人沿工件坐标系的 Y 轴（Y_W）正向运动。按下 [-Y（J2）] 键，机器人沿工件坐标系的 Y 轴（Y_W）负向运动。

按下 [+Z（J2）] 键，机器人沿工件坐标系的 Z 轴（Z_W）正向运动。按下 [-Z（J2）] 键，机器人沿工件坐标系的 Z 轴（Z_W）负向运动。

（3）Changing the flange surface posture　改变法兰面方位

The position of the control point does not change. Change the direction of the flange in accordance with the work coordinate system.

本操作不改变控制点的位置，只按工件坐标系改变法兰面的方位（图 4-45）。

When the [+A（J4）] key is pressed, the X axis will rotate in the plus direction of the work coordinate system. When the [-A（J4）] key is pressed, rotate in the minus direction.

When the[+B（J5）] key is pressed, the Y axis will rotate in the plus direction of the work coordinate system. When the [-B（J5）] key is pressed, rotate in the minus direction.

When the [+C（J6）] key is pressed, the Z axis will rotate in the plus direction of the work coordinate system. When the [-C（J6）] key is pressed, rotate in the minus direction.

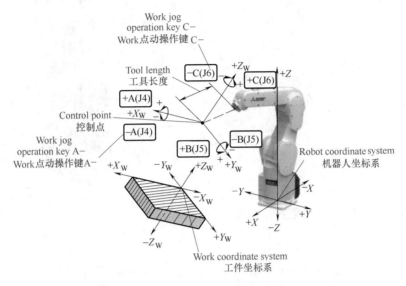

图 4-45　垂直型机器人在工件坐标系的点动（改变法兰面方位）

按下 [+A（J4）] 键，机器人按工件坐标系绕 X 轴正向旋转。按下 [-A（J4）] 键，机器人按工件坐标系绕 X 轴负向旋转。

按下 [+B（J5）] 键，机器人按工件坐标系绕 Y 轴正向旋转。按下 [-B（J5）] 键，机器人按工件坐标系绕 Y 轴负向旋转。

按下 [+C（J5）] 键，机器人按工件坐标系绕 Z 轴正向旋转。按下 [-C（J5）] 键，机器人按工件坐标系绕 Z 轴负向旋转。

(4) Move around the X axis　绕 X 轴点动（图 4-46）

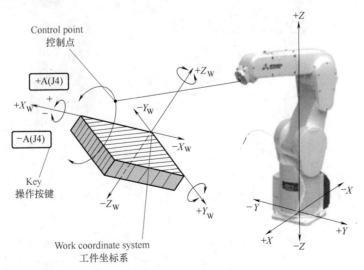

图 4-46　垂直型机器人在 Ex-T 坐标系的绕 X 轴点动

(5) Move around the Y axis　绕 Y 轴点动（图 4-47）

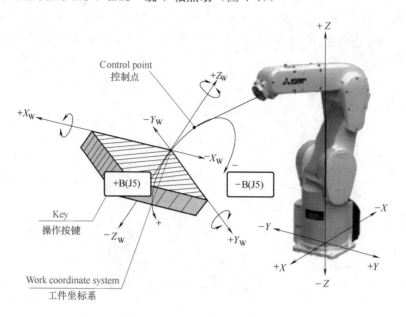

图 4-47　垂直型机器人在 Ex-T 坐标系的绕 Y 轴点动

(6) Move around the Z axis　绕 Z 轴点动（图 4-48）

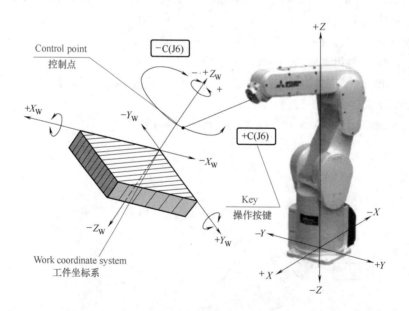

图 4-48　垂直型机器人在 Ex-T 坐标系绕 Z 轴点动

When the [+A (J4)] key is pressed, the control point will rotate in the plus direction around the X axis (X_W) of work coordinate system (Ex-T coordinate system). When the [-A (J4)] key is pressed, the control point will rotate in the minus direction.

When the [+B (J5)] key is pressed, the control point will rotate in the plus direction around the Y axis (Y_W) of work coordinate system (Ex-T coordinate system). When the [−B (J5)] key is pressed, the control point will rotate in the minus direction.

When the [+C (J6)] key is pressed, the control point will rotate in the plus direction around the Z axis (Z_W) of work coordinate system (Ex-T coordinate system). When the [−C (J6)] key is pressed, the control point will rotate in the minus direction.

按下 [+A（J4）] 键，控制点按工件坐标系（Ex-T 坐标系）绕 X 轴正向旋转。按下 [−A（J4）] 键，控制点按工件坐标系（Ex-T 坐标系）绕 X 轴负向旋转。

按下 [+B（J5）] 键，控制点按工件坐标系（Ex-T 坐标系）绕 Y 轴正向旋转。按下 [−B（J5）] 键，控制点按工件坐标系（Ex-T 坐标系）绕 Y 轴负向旋转。

按下 [+C（J6）] 键，控制点按工件坐标系（Ex-T 坐标系）绕 Z 轴正向旋转。按下 [−C（J6）] 键，控制点按工件坐标系（Ex-T 坐标系）绕 Z 轴负向旋转。

4.4 Aligning the hand　机械手整列操作

The posture of the hand attached to the robot can be aligned in units of 90 degrees.

机器人抓手的形位，可以按照 90° 的位置做整列动作。

This feature moves the robot to the position where the A, B and C components of the current position are set at the closest values in units of 90 degrees.

执行"整列动作"时，机器人会移动到当前位置的 A、B、C 成分中离 90° 位置最近的值。

The "aligning" is to make the robot's hand return to the nearest 90° from the "current position" (Fig. 4-49). In practice, this is a quick method if the hand is needed to quickly align the workpiece.

"整列功能"就是使机器人的抓手就近回到距离"当前位置"最靠近的 90° 位置（图 4-49）。在实际应用中，如果需要抓手迅速对准工件，这是一种快捷的方法。

1）If the "TOOL coordinate system" is not set, the hand will reach the previous position in Fig. 4-49 after passing through the "aligning".

如果没有设置"TOOL 坐标系"，经过"整列"后，抓手就到达图 4-49 中的前一个位置。

2）If the "TOOL coordinate system" is set, the hand will reach the latter position in Fig. 4-49 after passing through the "aligning".

如果设置"TOOL 坐标系"，经过"整列"后，抓手就到达图 4-49 中的后一个位置。

You can see it's based on control points. The position of the "control point" remains the same, while the position of the hand changes.

可以看到是以"控制点"为基准。"控制点"位置不变，抓手的位置发生改变。

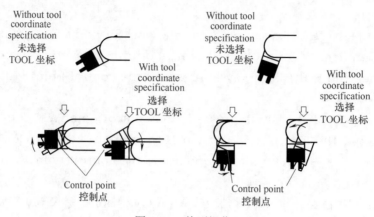

图 4-49 整列操作

Chapter 5
Programming
机器人编程

5.1 Creation procedures 编制程序的流程

(1) Flowchart 流程图 (图 5-1)

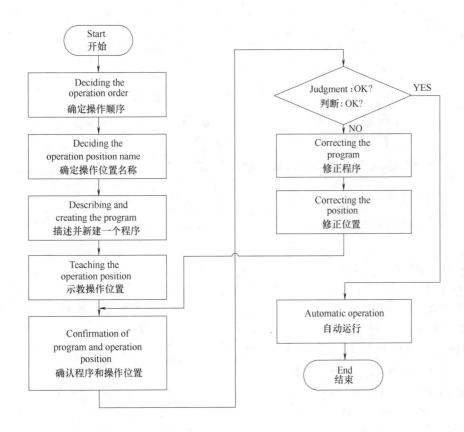

图 5-1 编制程序流程

（2）Robot's work　机器人的工作过程（图 5-2）

图 5-2　机器人工作过程样例

5.2　Creating the program　新建程序

（1）Deciding the operation order　确定机器人操作顺序（图 5-3）

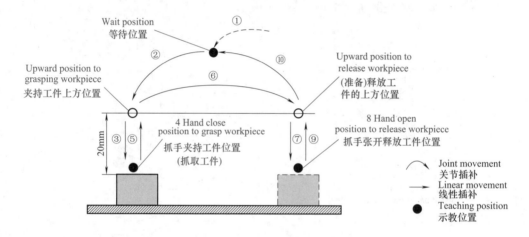

图 5-3　确定机器人操作顺序

机器人操作顺序：

① Wait position——到达等待位置；

② Upward position to grasping workpiece——移动到抓取工件的上方位置；

③ Move down——下降；

④ Hand close position to grasp workpiece——抓取工件；

⑤ Upward movement——上行；

⑥ Upward position to release workpiece——移动到放置工件上方位置；

⑦ Move down——下降；

⑧ Hand open position to release workpiece——抓手张开释放工件；

⑨ Upward movement——上行；

⑩ Return to wait position——回到等待位置。

(2) Deciding the operation position name 确定操作位置名称（图 5-4、表 5-1）

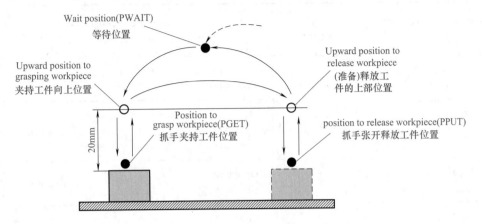

图 5-4 确定机器人操作位置名称

表 5-1 操作位置名称信息

Name 名称	Position variable name 位置变量名	Teaching 示教	Remarks 备注
Wait position 等待位置	PWAIT	Required 需要	
Upward position to grasping workpiece 夹持工件上部位置		Not required 不需要	Designate with commands 使用指令编程
Position to grasp workpiece 夹持位置	PGET	Required 需要	
Upward position to release workpiece 释放工件上部位置		Not required 不需要	Designate with commands 使用指令编程
Position to release workpiece 释放工件位置	PPUT	Required 需要	

(3) Describing and creating the program 描述并编制程序（表 5-2）

表 5-2 程序操作指令与样例

Target operation and work 操作	Command 指令	Example of designation 编程样例	
Joint movement 关节插补	Mov	Move to position variable PWAIT 移动到 PWAIT 点	Mov PWAIT
		Move to 20mm upward position variable PGET 移动到 PGET 点上部 20mm	Mov PGET,+20

续表

Target operation and work 操作	Command 指令	Example of designation 编程样例	
Linear movement 直线插补	Mvs	Move to position variable PGET 移动到 PGET 点	Mvs PGET
		Move to 20mm upward position variable PGET 移动到 PGET 点上部 20mm	Mvs PGET,+20
Hand open 张开抓手	Hopen	Open hand 1 张开抓手 1	Hopen 1
Hand close 抓手夹持	Hclose	Close hand 1 抓手 1 夹持	Hclose 1
Wait 等待	Dly	Wait 1 second 等待 1 秒	Dly 1.0
End 结束	End	End the program 程序结束	End

(4) Convert to program instruction，转换成程序指令

```
1   Mov PWAIT         Move to wait position（joint movement）
2   Mov PGET,+20      Move to 20mm upward workpiece（joint movement）
3   MVS PGET          Move to position to grasp workpiece（linear movement）
4   HClose 1          Grasp workpiece（hand close）
5   Dly 1.0           Waits for 1 seconds
6   MVS PGET,+20      Move 20mm upward（linear movement）
7   Mov PPUT,+20      Move to 20mm upward position to release workpiece（joint movement）
8   MVS PPUT          Move to position to place workpiece（linear movement）
9   HOpen 1           Release workpiece（hand open）
10  Dly 1.0           Waits for 1 seconds
11  MVS PPUT,+20      Move 20mm upward（linear movement）
12  Mov PWAIT         Move to wait position（joint movement）
13  End               End

1   Mov PWAIT         移动到等待点"PWAIT"。
```

2	Mov PGET,+20	移动到"PEGT"点上部 +20mm 处。
3	MVS PGET	移动到"PEGT"点。
4	HClose 1	抓手夹持工件。
5	Dly 1.0	暂停 1.0 秒。
6	MVS PGET,+20	移动到"PEGT"点上部 +20mm 处。
7	Mov PPUT,+20	移动到"PPUT"点上部 +20mm 处。
8	MVS PPUT	移动到"PPUT"点。
9	HOpen 1	打开抓手。
10	Dly 1.0	暂停 1.0 秒。
11	MVS PPUT,+20	移动到"PPUT"点上部 +20mm 处。
12	Mov PWAIT	移动到等待点"PWAIT"。
13	End	结束。

Chapter 6
Commands of the Robot
机器人编程指令

6.1 Coordinate system description of the robot 机器人使用的坐标系种类

（1）The robot's coordinate system　机器人坐标系（图6-1）

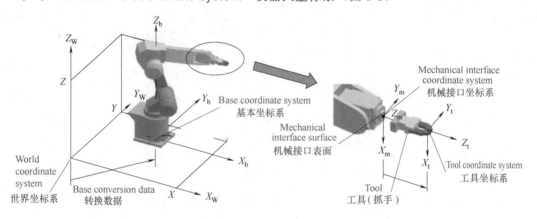

图6-1　机器人坐标系

Explaination　说明：

The robot's coordinate system has following four.

机器人坐标系有以下4种。

1）World coordinate system: the coordinate system as the standard for displaying the current position of robot.

世界坐标系：由 X_W-Y_W-Z_W 构成，是表示机器人"当前位置"的基准坐标系。

2）Base coordinate system: origin is J1 axis rotation center on the bottom of the robot. A coordinate system established with reference to the robot mounting surface. It is set by specifying parameter MEXBS with data on a center position for robot installation (base conversion data) as viewed from the world coordinate system or by executing a base command. By default, because the base conversion data is set to zero (0), the world coordinate system is in agreement with the base coordinate system.

基本坐标系：基本坐标系是以机器人的基座安装面（原点位于机器人底面的 J1 轴旋转中心）为基准确定的坐标系，由 X_b-Y_b-Z_b 构成。

世界坐标系与基本坐标系的相互关系可以通过设置参数 MEXBS 或执行 Base 指令进行设定。初始设定时，MEXBS=0，因此世界坐标系和基本坐标系一致。

3) Mechanical interface coordinate system: origin is J6 axis rotation center on the tool installation surface. A coordinate system established with reference to the robot's mechanical interface.

机械接口坐标系：机械接口坐标系是以 J6 轴旋转中心和安装抓手的法兰面为基准确定的坐标系，由 X_m-Y_m-Z_m 构成。

4) Tool coordinate system: a coordinate system established with reference to the robot's mechanical interface. Its relation to the interface coordinate system is determined by the tool data (i.e., by specified settings for parameter MEXTL or by the execution of a tool command).

工具坐标系：由安装在机器人法兰面上的"工具（抓手）"确定的坐标系。由 X_t-Y_t-Z_t 构成，工具坐标系与机械 I/F 坐标系之间的关系取决于工具数据（可由参数 MEXTL 设定，或执行 Tool 指令后确定工具坐标系）。

（2）Base conversion （坐标系）基本转换（图 6-2）

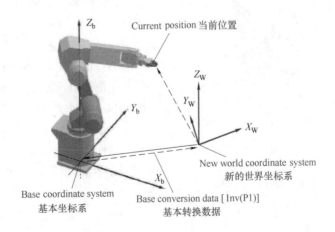

图 6-2　坐标系转换

Explaination: P1 is teaching position data.

说明：P1 是示教位置数据。

（3）Position data　位置数据（图 6-3）

Position data for the robot is comprised of six elements which indicate the position of the hand's leading end (mechanical interface center where no tool setting is made) (X, Y, and Z) and the robot's posture (A, B, and C), plus a structure flag. Each element constitutes reference data for the robot's world coordinate system.

机器人的位置数据是指"抓手工作点"(未进行工具坐标系设定时为机械 I/F 中心)的位置(X, Y, Z)与"立体形位"角度(A, B, C)这 6 个数据以及结构标志的数据。各数据的值均以世界坐标系为基准。

Meaning:

X, Y, Z: Coordinate data. Position of the robot hand's leading end (in mm).

A, B, C: Posture data. Angle that defines the robot's posture (in degrees).

A: Angle of rotation on X axis.

B: Angle of rotation on Y axis.

C: Angle of rotation on Z axis.

含义:

X, Y, Z: 坐标数据。即机器人的抓手工作点在世界坐标系中的位置(单位为 mm)。

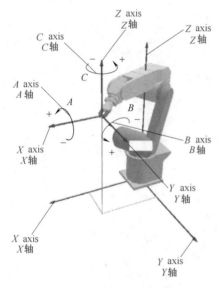

图 6-3 "立体形位"角度位置数据

A, B, C: "立体形位"数据。表示机器人"立体形位"的角度(单位为"°")。

A: 绕 X 轴旋转的角度。

B: 绕 Y 轴旋转的角度。

C: 绕 Z 轴旋转的角度。

A, B, and C represent the robot's posture in the coordinate system of its hand's leading end (or flange center where no tool setting is made), each indicating a angle of rotation on the X axis, Y axis, and Z axis of the world coordinate system. Rotation corresponding to the direction of a right-handed screw when you look at the "+" side of each coordinate axis is "+" rotation. Also, rotation is set to take place in a predetermined sequence, and the amount of rotation is calculated (controlled) first for a rotation on the Z axis, followed by one on the Y axis and one on the Z axis in the order shown.

A, B, C 表示"抓手工作点"(未进行工具坐标系设定时为法兰中心)的"立体形位",它们分别表示"抓手工作点"绕世界坐标系的 X 轴、Y 轴、Z 轴旋转的角度。旋转方向以各坐标轴"+"向为基准,右转方向为"+"方向(右手法则)。旋转顺序,按照 Z 轴旋转→Y 轴旋转→X 轴旋转的顺序旋转并进行计算(控制)。

6.2 Mechanical interface coordinate system
机械接口坐标系

(1) About mechanical interface coordinate system 机械接口坐标系定义

A coordinate system having its origin point chosen at the center of the flange is called a mechanical interface (I/F) coordinate system. X axis, Y axis and Z axis of the mechanical interface coordinate system are denoted as X_m, Y_m and Z_m, respectively, as shown in Fig. 6-4.

以法兰中心为原点的坐标系被称为"机械 I/F（接口）坐标系"。机械 I/F 坐标系的 X 轴、Y 轴、Z 轴用 X_m、Y_m、Z_m 表示，见图 6-4。

Z_m is an axis which passes through the flange center and is perpendicular to the flange face. The direction which goes outside from the flange face is "+" (plus). X_m and Y_m are coplanar with the flange face. A line joining the flange center with the positioning pin hole is represented by X_m axis. "+" direction of the X_m axis is opposite to the pin hole as seen from the center.

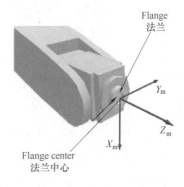

Z_m 轴为穿过法兰中心而且垂直于法兰面的轴。从法兰面向外的方向为"+"方向。X_m、Y_m 被设定在法兰面内。法兰中心和定位销孔的连线即为 X_m 轴。X_m 轴的"+"方向为从中心到销孔相反方向。

图 6-4　机械 I/F 坐标系

(2) Flange rotates 法兰旋转

When the flange rotates, the mechanical interface coordinate system also rotates (Fig. 6-5).

法兰转动，机械接口（I/F）坐标系亦随之转动（图 6-5）。

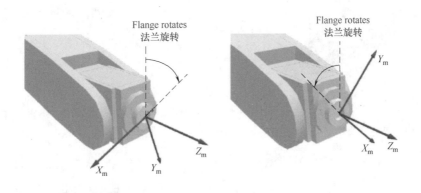

图 6-5　法兰旋转与机械接口坐标系

6.3 Tool coordinate system 工具坐标系

6.3.1 Definition 定义

A tool coordinate system is one that is defined for the leading end of the robot hand (control point for the robot hand). It is obtained by shifting the origin point of a mechanical interface coordinate system to the leading end of the robot hand (control point hand) and adding given rotational elements.

X axis, Y axis and Z axis of the tool coordinate system are denoted as X_t, Y_t and Z_t, respectively (Fig.6-6).

TOOL 坐标系是以"抓手工作点"为基准的坐标系，它将机械接口坐标系的原点移到"抓手工作点"，并增加了旋转的角度。

TOOL 坐标系的 X 轴、Y 轴、Z 轴用 X_t、Y_t、Z_t 表示（图6-6）。

Tool data consist of the same elements as position data.

X, Y, Z: Amount of shift. Amount by which the origin point of the mechanical interface coordinate system is shifted to agree with that of the tool coordinate system (in mm).

A, B, C: Angle of rotation of each coordinate axis (in degrees).

A: Angle of rotation on X axis.

B: Angle of rotation on Y axis.

C: Angle of rotation on Z axis.

TOOL 数据具有与位置数据相同的内容。

X, Y, Z：移动量（单位为 mm）。

A, B, C：绕坐标轴的旋转角度（单位为"°"）。

A：绕 X 轴的旋转角度。

B：绕 Y 的旋转角度。

C：绕 Z 的旋转角度。

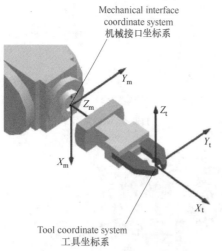

图6-6　机械界面坐标系与工具坐标系

6.3.2 Effects of using tool coordinate system 使用工具坐标系的效果

(1) Comparison of motion in different coordinate systems 在不同坐标系中的运动比较（图 6-7）

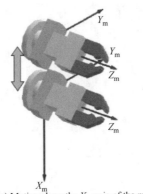
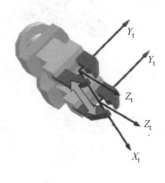

(a) Motion along the X_m axis of the mechanical interface coordinate system
在机械接口坐标系中沿X_m轴运动

(b) Motion along the X_t axis of the tool coordinate system
在TOOL坐标系中沿X_t轴运动

图 6-7　使用工具坐标系的点动操作比较

(2) Comparison of jog operations using/without tool coordinate systems 使用 / 不使用工具坐标系的点动操作比较（图 6-8）

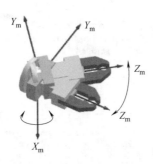
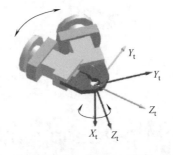

(a) The robot hand rotates on the X_m axis of the mechanical interface coordinate system, thus having a wide range of motion at its leading end
绕机械接口坐标系的X_m轴旋转，抓手前端大幅移动

(b) The robot hand rotates on the X_t axis of the tool coordinate system. Rotational motion on the leading end of the robot hand permits a change of posture without the need to displace the work from its original position
绕TOOL坐标系的X_t轴旋转。以抓手前端为中心的旋转，可在不偏离工件位置的情况下变更3D形位

图 6-8　使用 / 不使用工具坐标系的点动操作比较

(3) Approach and pull out motion　接近与拉出操作（图6-9）

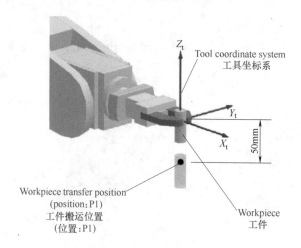

图6-9　接近与拉出操作

(4) Rotational motion in tool coordinate system　在工具坐标系中的旋转运动（图6-10）

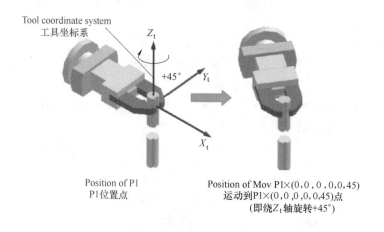

图6-10　在工具坐标系中的旋转运动

6.4　Robot operation control　机器人动作指令

6.4.1　Interpolation movement　插补指令

(1) Joint interpolation movement Mov（Move）　关节插补指令 Mov（图6-11）

Explaination: Using joint interpolation operation, robot moves from the current position to the destination position.

说明：使用"关节插补"操作，机器人从"当前位置"运动到"目标位置"。

The joint angle differences of each axis are evenly interpolated at the starting point and endpoint positions. This means that the path of the tip cannot be guaranteed.

在从起点向终点的插补运动中，每一轴的关节旋转角度是不同的。因此机器人的运行轨迹无法确定。

(2) Linear interpolation movement Mvs (Move S)　直线插补指令 Mvs

Function: Carries out linear interpolation movement from the current position to the movement target position. (Fig.6-12)

功能：从"当前位置"向"目标位置"做直线插补运动（图 6-12）。

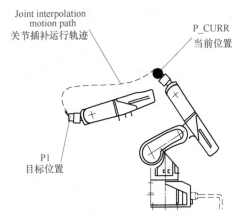

图 6-11　关节插补运行轨迹示意图

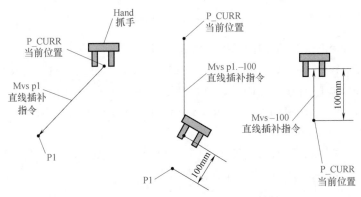

图 6-12　Mvs 插补运行轨迹示意图

(3) Mvc　三维真圆插补指令

Carries out 3D circular interpolation in the order of start point P1, transit point P2, transit point P3 and start point (Fig.6-13).

从起点开始，按起点 P1、通过点 P2、通过点 P3、起点 P1 的顺序进行三维真圆插补（图 6-13）。

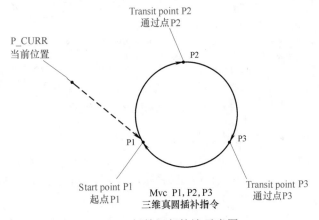

图 6-13　Mvc 插补运行轨迹示意图

（4）Mvr　三维圆弧插补指令（图6-14）

Function: Carries out 3-dimensional circular interpolation movement from the start point to the end point via transit points.

功能：从起点开始，经由"通过点"到"终点"，执行三维圆弧插补。

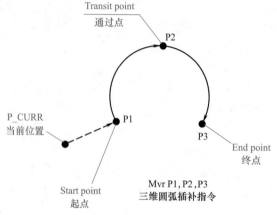

图6-14　Mvr插补运行轨迹图

（5）Mvr2（Move R 2）　三维圆弧插补指令（图6-15）

Function: Carries out 3-dimensional circular interpolation motion from the start point to the end point on the arc composed of the start point, end point, and reference points. The direction of movement is in a direction that does not pass through the reference points.

功能：由起点、终点、参考点确定一圆弧，从起点到终点，执行三维圆弧插补。移动方向为不经过参考点的方向（即移动轨迹不经过参考点）。

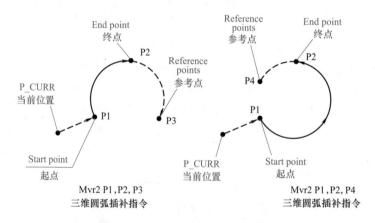

图6-15　Mvr2插补运行轨迹示意图

（6）Mvr3（Move R 3）　三维圆弧插补指令（图6-16）

Function: Carries out 3-dimensional circular interpolation movement from the start point to the end point on the arc composed of the center point, start point and end point.

功能：由起点、终点、中心点构成一圆弧，执行从起点到终点的三维圆弧插补。

Chapter 6 Commands of the Robot

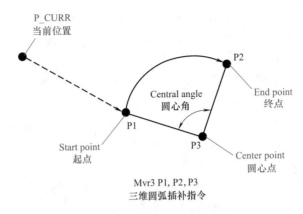

图 6-16 Mvr3 插补运行轨迹示意图

6.4.2 Cnt（Continuous）连续轨迹运行

(1) Function 功能（图 6-11）

Designates continuous movement control for interpolation. Shortening of the operating time can be performed by carrying out continuous movement.

指定执行"连续插补"。通过执行连续插补运动控制，可缩短运行时间。

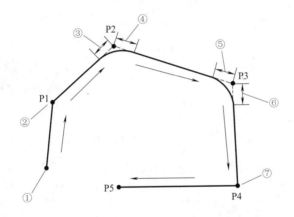

图 6-17 连续轨迹运行示意图

① Start position of movement——起点。
② It moves to P1 first and then to P2 since continuous operation is not set up. 因为没有设置"连续功能"，所以先运动到 P1 再运动到 P2。
③ End side neighborhood distance——结束侧接近距离。
④ Starting side neighborhood distance——启动侧接近距离。
Continuous operation is performed at a distance shorter than the smaller of the neighborhood distance（the initial setting value in the robot controller）when moving to P2 and the fulcrum neighborhood point（100 mm）when moving to P3.

做连续轨迹运行。过渡圆弧见图 6-17，过渡圆弧的构成值为小于③、④的数值。

⑤ End side neighborhood distance——结束侧接近距离。
⑥ Starting side neighborhood distance——启动侧接近距离。

Continuous operation is performed at a distance shorter than the smaller of the neighborhood distance (200mm) when moving to P3 and the fulcrum neighborhood point (300mm) when moving to P4.

做连续轨迹运行。过渡圆弧见图6-17，过渡圆弧的构成值为小于⑤（200mm）、⑥（300mm）的数值。

⑦ Although the neighborhood distance (300mm) when moving to P4 has been set, continuous operation when moving to P5 has been canceled. Therefore, it moves to P4 first, and then moves to P5.

虽然已经设置向P4运动的"启动侧接近距离=300mm"，但向P5点移动已经解除了"连续轨迹运行"，所以运行轨迹是先移动到P4，再移动到P5点。

(2) Comparison of continuous operation and "point-to-point" operation mode 连续运行与"点到点"运行方式的比较（图6-18）

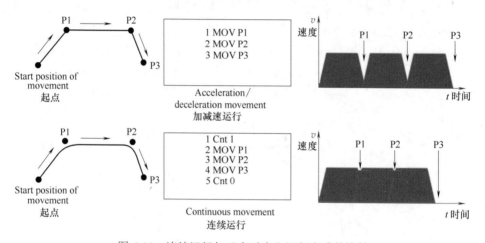

图6-18 连续运行与"点对点"运行方式的比较

1) 在图6-18中，"加减速模式"运行，运动程序为：

```
1 MOV P1
2 MVS P2
3 MOV P3
```

It decelerates and accelerates to P1 and P3. After moving to the target point moves to the next target position.

在从P1到P3的运行中，是加速和减速的运行模式，在到达一个目标点后，再运行到下一目标点。

2) 在"连续运行模式"中：

```
1 Cnt1
2 MOV P1
3 MVS P2
4 MOV P3
5 Cnt 0
```

It passes through the neighborhood of P1 and P2, and then moves to P3.

以圆弧轨迹通过 P1、P2 的邻近位置，再运动到 P3 点。

6.4.3 Pallet operation 码垛指令

（1）Function 功能

"Pallet 指令"也翻译为"托盘指令"或"码垛指令"，实际上是一个计算矩阵方格中各"点位中心（位置）"的指令，该指令需要设置"矩阵方格"有几行几列、起点终点、对角点位置、计数方向。由于该指令通常用于码垛动作，所以也就被称为"码垛指令"。

（2）Instruction format 指令格式

Def Plt——定义"托盘结构"指令（定义一个矩阵结构）。

Def Plt< 托盘号 >< 起点 >< 终点 A >< 终点 B >[< 对角点 >]< 列数 a >< 行数 b >< 托盘类型 >

Plt——指定托盘中的某一点。

（3）Sample instruction 指令样例

如图 6-19 所示。

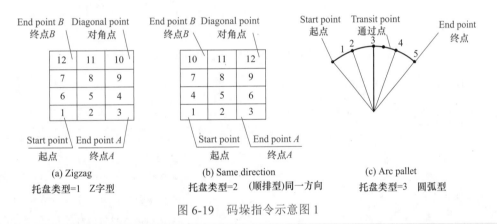

图 6-19 码垛指令示意图 1

```
1 Def Plt 1, P1, P2, P3, 3, 4, 1        '——3 点型托盘定义指令。
2 Def Plt 1, P1, P2, P3, P4, 3, 4, 1    '——4 点型托盘定义指令。
```

3 点型托盘定义指令——指令中只给出起点、终点 A、终点 B。

4 点型托盘定义指令——指令中给出起点、终点 A、终点 B、对角点。

（4）Explaination 说明

1）托盘号　可以将一个矩阵视作一个"托盘"（因为实际工程中，工件摆放在一个托盘上），系统可设置 8 个托盘。本数据设置第几号托盘。

2）起点／终点／对角点　如图 6-19 所示，用"位置点"设置。

3）< 列数 a >　起点与终点 A 之间列数。

4）< 行数 b >　起点与终点 B 之间行数。

5）< 托盘类型 >　设置托盘中"各位置点"分布类型。

托盘类型 =1——Z 字型。

托盘类型 =2——顺排型。

托盘类型 =3——圆弧型。

托盘类型 =11——Z 字型。

托盘类型 =12——顺排型。

托盘类型 =13——圆弧型。

（5）Sample program　程序样例（图 6-20）

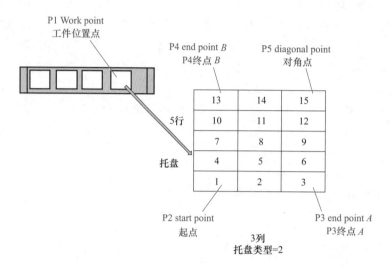

图 6-20　码垛指令示意图 2

Sample program 程序样例：

10 Def Plt 1, P2, P3, P4, P5, 3, 5, 2	'——Define the pallet. Pallet No.=1, start point=P2, end point A=P3, end point B=P4, diagonal point=P5, quantity a=3, quantity b=5, pallet pattern=2（same direction）./ 设定托盘为 1 号托盘，起点 =P2，终点 A=P3，终点 B=P4，对角点 =P5，列数 a= 3，行数 b=5，顺排型。
11 M1=1	'——Substitute value 1 in numeric variable M1（M1 is used as a counter）./ 设置变量 M 1 =1。
12 *LOOP	'——Designate label LOOP at the jump destination./ 在跳跃目标处设置循环指令标志。
13 Mov P1, -50 *1	'——Move with joint interpolation from P1 to a position retracted 50mm in hand direction./ 关节插补运行到辅助工作点，辅助工作点

Chapter 6　Commands of the Robot

	沿抓手方向距离 P1 点 50mm。
14 Ovrd 50	'——Set movement speed to half of the maximum speed./ 设置运行速度为最大速度的一半。
15 Mvs P1	'——Move linearly to P1 (go to grasp workpiece)./ 直线插补到 P1 点（准备夹持工件）。
16 HClose 1	'——Close hand 1. (grasp workpiece)./1 号抓手闭合（夹持工件）。
17 Dly 0.5	'——Wait 0.5 seconds./ 等待 0.5s。
18 Ovrd 100	'——Set movement speed to maximum speed./ 设置运行速度为最大速度。
19 Mvs, -50 *1	'——Move linearly from current position (P1) to a position retracted 50mm in hand direction (lift up workpiece)./ 从当前位置 P1 点关节插补运行到辅助工作点，辅助工作点沿抓手方向距离 P1 点 50mm（提起工件）。
20 P10=(Plt1, M1)	'——Operate the position in pallet No.1 indicated by the numeric variable M1, and substitute the results in P10./ 设置 P10 点=1 号托盘中的 M1 点。
21 Mov P10, -50 *1	'——Move with joint interpolation from P10 to a position retracted 50mm in hand direction./ 关节插补运行到辅助工作点，辅助工作点沿抓手方向距离 P10 点 50mm。
22 Ovrd 50	'——Set movement speed to half of the maximum speed. / 设置运行速度为最大速度的一半。
23 Mvs P10	'——Move linearly to P10 (go to place workpiece)./ 直线插补运行到 P10 点（准备放下工件）。
24 HOpen 1	'——Open hand 1. (place workpiece)./ 打开 1 号抓手（放

```
25 Dly 0.5                              '——Wait 0.5 seconds./ 等待 0.5s。
26 Ovrd 100                             '——Set movement speed to
                                          maximum speed./ 设置运行速度
                                          为最大速度。
27 Mvs, -50                             '——Move linearly from current
                                          position (P10) to a position
                                          retracted 50mm in hand
                                          direction. (separate from
                                          workpiece)./ 从当前位置 P10 点
                                          直线插补运行到辅助工作点，辅
                                          助工作点沿抓手方向距离 P10 点
                                          50mm（离开工件）。
28 M1=M1+1                              '——Increment numeric variable
                                          M1 by 1. (advance the pallet
                                          counter)./ 设置变量 M1=M1+1。
29 If M1<=15 Then *LOOP                 '——If numeric variable M1 value
                                          is less than 15, jump to label
                                          LOOP and repeat process. If
                                          more than 15, go to next step./
                                          判断：如果 M1 ≤ 15，则跳转到
                                          标记有 *LOOP 的程序行。如果
                                          M1 大于 15，就执行下一行。
30 End                                  '——End the program./ 结束程序。
```

6.5　Robot operation control　机器人运行控制

（1）Joint interpolation movement　关节插补运行

The robot moves with joint axis unit interpolation to the designated position (the robot interpolates with a joint axis unit, so the end path is irrelevant).

机器人以关节轴旋转角度为单位，插补运行到设定位置（由于机器人以角度单位进行插补运行，所以机器人的运行轨迹是无法描述的）。

Command word: Mov.

指令字：Mov。

Explanation 说明：

The robot moves to the designated position with joint interpolation. An appended statement Wth or WthIf can be designated.

机器人以关节插补方式运行到设定位置。在运行过程中其附加动作可以使用附加语句进行设置。如图 6-21。

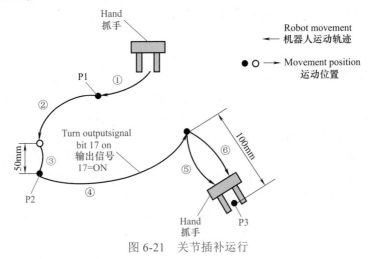

图 6-21 关节插补运行

Program explanation 程序说明：

```
1 Mov P1                          '——① Move to P1./ 运动到 P1 点。
2 Mov P2, -50                     '——② Move from P2 to a position
                                  retracted 50mm in the hand
                                  direction./ 运动到辅助工作
                                  点，该辅助工作点在 P2 点上方
                                  50mm，方向沿抓手方向。
3 Mov P2                          '——③ Move to P2./ 运动到 P2 点。
4 Mov P3, -100 Wth M_Out (17) =1  '——④ Start movement from
                                  P3 to a position retracted
                                  100mm in the hand
                                  direction, and turn ON
                                  output signal bit 17./ 从 P2
                                  点运动到辅助工作点，该辅助
                                  工作点在 P3 点上方 100mm，
                                  方向沿抓手方向。
5 Mov P3                          '——⑤ Moves to P3./ 运动到 P3 点。
6 Mov P3, -100                    '——⑥ Return from P3 to a
                                  position retracted 100mm in
                                  the hand direction./ 从 P3 点返
                                  回到辅助工作点，该辅助工作点
                                  在 P3 点上方 100mm，方向沿
                                  抓手方向。
7 End                             '——End the program./ 程序结束。
```

(2) Linear interpolation movement　直线插补运行

The end of the hand is moved with linear interpolation to the designated position.

抓手工作点以直线插补方式运动到设定位置。

Command word: MVS.

指令字：MVS。

Explanation 说明：

The robot moves to the designated position with linear interpolation. It is possible to specify the interpolation form using the TYPE instruction. An appended statement Wth or WthIf can be designated.

机器人以直线插补方式运行到设定位置。因此可以使用TYPE型指令设定这种插补动作。在运行过程中其附加动作可以使用附加语句进行设置（图6-22）。

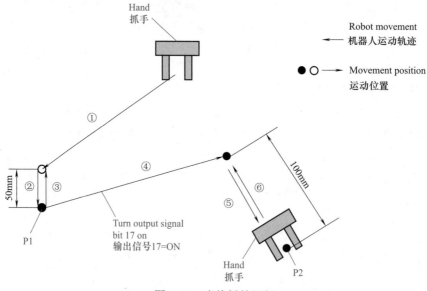

图6-22　直线插补运行

Program explanation 程序说明：

1 Mvs P1, -50	'——① Move with linear interpolation from P1 to a position retracted 50mm in the hand direction./ 以直线插补方式运动到辅助工作点，该辅助工作点在P1点上方50mm，方向沿抓手方向。
2 Mvs P1	'——② Move to P1 with linear interpolation./ 以直线插补方式运动到P1点。
3 Mvs, -50	'——③ Move with linear interpolation from the current position (P1) to a position retracted 50mm in the hand direction./ 从P1点以直线插补运动到辅助工作点，该辅助工作点在P1点上方50mm，方向沿抓手方向。

```
4 Mvs P2, -100 Wth M_Out (17) =1    '──④ Output signal bit 17 is turned on
                                        at the same time as the robot starts
                                        moving./ 以直线插补运动到辅助工作点,
                                        该辅助工作点在 P2 点上方 100mm, 方向
                                        沿抓手方向。同时指令输出信号 17=ON。

5 Mvs P2                            '──⑤ Move with linear interpolation to
                                        P2./ 以直线插补方式运动到 P2 点。

6 Mvs, -100                         '──⑥ Move with linear interpolation
                                        from the current position (P2) to a
                                        position retracted 50mm in the hand
                                        direction./ 以直线插补运动到辅助工作点,
                                        该辅助工作点在 P2 点上方 100mm, 方向
                                        沿抓手方向。

7 End                               '──End the program./ 程序结束。
```

(3) Circular interpolation movement   圆弧插补（图 6-23）

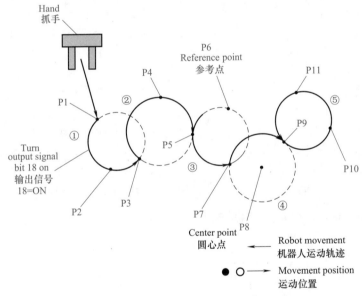

图 6-23 圆弧插补

Program Explanation   程序说明：

```
1 Mvr P1, P2, P3 Wth M_Out (18) =1  '──① Move between P1, P2, P3 as
```
an arc. The robot current position before movement is separated from the start point, so first the robot will move with linear operation to the start point (P1).Output signal bit 18 turns ON simultaneously with

the start of circular movement./ 对由P1、P2、P3点构成的圆弧进行插补运行。机器人当前位置在启动之前与起点是分开的，所以机器人首先是直线插补运行到起点，同时在机器人从起点开始圆弧插补后，输出信号18=ON。

```
2 Mvr P3, P4, P5              '——② Move between P3, P4, P5 as an arc./
                                 对P3、P4、P5构成的圆弧进行插补运行。

3 Mvr2 P5, P7, P6             '——③ Move as an arc over the circumference
                                 on which the start point (P5), reference
                                 point (P6) and end point (P7) in the
                                 direction that the reference point is not
                                 passed between the start point and end
                                 point./ 执行圆弧插补。圆弧由起点P5，参
                                 考点P6和终点P7构成。但经过的圆弧轨
                                 迹只是由P5和P7构成的一段。

4 Mvr3 P7, P9, P8             '——④ Move as an arc from the start
                                 point to the end point along the
                                 circumference on which the center
                                 point (P8), start point (P7) and
                                 end point (P9) are designated./ 执
                                 行圆弧插补。圆弧由起点P7、圆心点
                                 P8和终点P9构成。

5 Mvc P9, P10, P11            '——⑤ Move between P9, P10, P11, P9
                                 as an arc (1 cycle operation)./ 执行真
                                 圆插补，真圆由P9、P10、P11、P9构成。

6 End                         '——End the program./ 程序结束。
```

（4）Continuous movement　连续轨迹运行（图6-24）

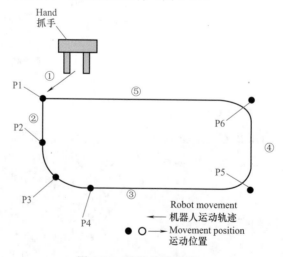

图6-24　连续轨迹运行

Program explanation   程序说明：

```
1 Mov P1              '——① Move with joint interpolation to P1./ 以关节插补运行
                         到 P1 点。
2 Cnt 1               '——Validates continuous movement (following movement
                         is continuous movement)./ 连续轨迹运行模式有效（以
                         下运行模式为"连续运行模式"）。
3 Mvr P2, P3, P4      '——② Move linearly to P2, and continuously moves to P4
                         with arc movement./ 直线插补到 P2 点并以圆弧模式连续
                         运行到 P4 点。
4 Mvs P5              '——After arc movement, move linearly to P5./ 在圆弧运动
                         之后，直线插补到 P5 点。
5 Cnt 1, 200, 100     '——③ Set the continuous movement's start point
                         neighborhood distance to 200mm, and the end point
                         neighborhood distance to 100mm./ 设置连续插补的起点
                         侧距离为 200mm，终点侧距离为 100mm。
6 Mvs P6              '——④ After moving to previous P5, move in succession
                         linearly to P6./ 在运行过 P5 点之后，再直线插补到 P6 点。
7 Mvs P1              '——⑤ Continuously moves to P1 with linear movement./
                         以直线插补方式连续运行到 P1 点。
8 Cnt 0               '——Invalidates the continuous movement./ 退出连续运行
                         模式。
9 End                 '——Ends the program./ 结束程序
```

（5）Acceleration/deceleration time and speed control   加减速时间及速度控制

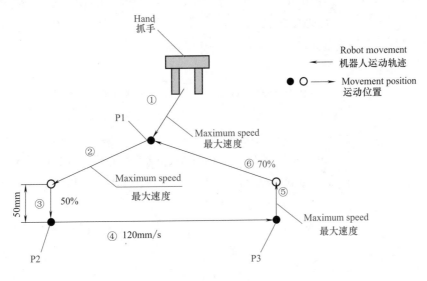

图 6-25   加减速时间和速度控制

Program explanation 程序说明：

1	Ovrd 100	'——Set the movement speed applied on the entire program to the maximum speed./ 设置速度倍率=100%，这是全部程序中的最大速度。
2	Mvs P1	'——① Move at maximum speed to P1./ 以最大速度运动至 P1 点。
3	Mvs P2, -50	'——② Move at maximum speed from P2 to position retracted 50mm in hand direction./ 以最大速度运动到辅助工作点，辅助工作点在 P2 点上方 50mm。
4	Ovrd 50	'——Set the movement speed applied on the entire program to half of the maximum speed./ 设置速度倍率=50%。
5	Mvs P2	'——③ Move linearly to P2 with a speed half of the default speed./ 以 50% 的速度倍率直线插补到 P2 点。
6	Spd 120	'——Set the end speed to 120mm/s（since the override is 50%, it actually moves at 60 mm/s）./ 设置速度为 120mm/s（因为速度倍率为 50%，所以实际速度为 60mm/s）。
7	Ovrd 100	'——Set the movement speed percentage to 100% to obtain the actual end speed of 120mm/s./ 设置速度倍率为 100%，则实际速度为 120mm/s。
8	Accel 70	'——Set the acceleration and deceleration to 70% of the maximum speed./ 设置加减速度为最大速度的 70%。
9	Mvs P3	'——④ Move linearly to P3 with the end speed 120mm/s./ 以 120mm/s 速度直线插补到 P3 点。
10	Spd M_NSpd	'——Return the end speed to the default value./ 设置速度等于初始速度值（缺省值）。
11	JOvrd 70	'——Set the speed for joint interpolation to 70%./ 设置关节插补速度（倍率）为 70%。
12	Accel	'——Return both the acceleration and deceleration to the maximum speed./ 设置加减速度到最大速度。
13	Mvs, -50	'——⑤ Move linearly with the default speed for linear movement from the current position (P3) to a position retracted 50mm in the hand direction./ 以缺省速度执行直线插补，从 P3 点运动到辅助工作点。
14	Mvs P1	'——⑥ Move to P1 at 70% of the maximum speed./ 以最大速度的 70% 移动到 P1 点。
15	End	'——Ends the program./ 程序结束。

Chapter 6 Commands of the Robot

(6) Confirming that the target position is reached 确认到达目标点（图 6-26）

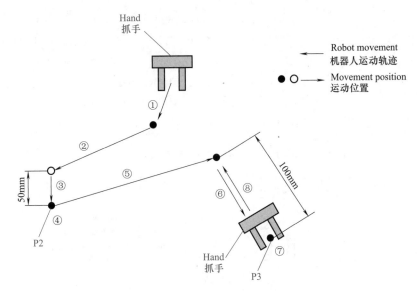

图 6-26 到达位置点的确认

Program explanation 程序说明：

1	Cnt 0	'——The Fine instruction is valid only when the Cnt instruction is OFF./ 仅仅在 Cnt 指令 =OFF 时，"Fine" 指令才有效。
2	Mvs P1	'——① Move with joint interpolation to P1./ 以直线插补模式运动到 P1 点。
3	Mvs P2, -50	'——② Move with joint interpolation from P2 to position retracted 50mm in hand direction./ 以直线插补模式运动到辅助工作点，辅助工作点距离 P2 点 50mm。
4	Fine 50	'——Set positioning finish pulse to 50./ 设置定位完成脉冲为 50。
5	Mvs P2	'——③ Move with linear interpolation to P2（Mvs completes if the positioning complete pulse count is 50 or less）./ 直线插补运动到 P2 点。如果定位脉冲计数等于或小于 50，就表示 MVS 指令执行完成。
6	M_Out (17) =1	'——④ Turn output signal 17 ON when positioning finish pulse reaches 50 pulses./ 当定位计数脉冲 =50，输出信号 17=ON。
7	Fine 1000	'——Set positioning finish pulse to 1000./ 设置定位完成脉冲为 1000。
8	Mvs P3, -100	'——⑤ Move linearly from P3 to position retracted 100mm in hand direction./ 以直线插补模式运动到辅助工作点，辅助工作点距离 P3 点 100mm。
9	Mvs P3	'——⑥ Move with linear interpolation to P3./ 以直线插补模式

运动到 P3 点.

```
10 Dly 0.1           '——Perform the positioning by the timer./ 暂停 0.1s。
11 M_Out(17)=0       '——⑦ Turn output signal 17 off./ 指令输出信号 17=OFF。
12 Mvs, -100         '——⑧ Move linearly from current position (P3) to position
                         retracted 100mm in hand direction./ 从 P3 点直线插补到辅
                         助工作点，辅助工作点距离 P3 点 100mm。
13 End               '——Ends the program./ 结束程序。
```

（7）High path accuracy control　高精度轨迹控制（图 6-27）

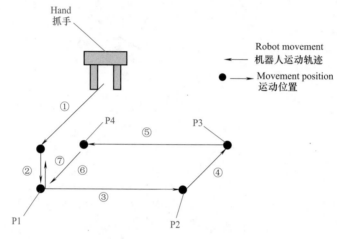

图 6-27　高精度轨迹控制

Program explanation 程序说明：

```
1 Mov P1, -50     '——① Move with joint interpolation from P1 to position
                      retracted 50mm in hand direction./ 关节插补到辅助工作点，
                      辅助工作点距离 P1 点 50mm。
2 Ovrd 50         '——Set the movement speed to half of the maximum speed./ 设
                      置运动速度为最大速度的一半。
3 Mvs P1          '——② Move with linear interpolation to P1./ 直线插补运动到 P1
                      点。
4 Prec On         '——The high path accuracy mode is enabled./ 开启"高精度轨迹
                      控制"模式。
5 Mvs P2          '——③ Move the robot from P1 to P2 with high path accuracy./
                      机器人以"高精度轨迹控制"模式从 P1 到 P2。
6 Mvs P3          '——④ Move the robot from P2 to P3 with high path accuracy./
                      机器人以"高精度轨迹控制"模式从 P2 到 P3。
7 Mvs P4          '——⑤ Move the robot from P3 to P4 with high path accuracy./
                      机器人以"高精度轨迹控制"模式从 P3 到 P4.。
8 Mvs P1          '——⑥ Move the robot from P4 to P1 with high path accuracy./
```

机器人以"高精度轨迹控制"模式从 P4 到 P1。

```
9  Prec Off         '——The high path accuracy mode is disable./ 关闭"高精度轨迹
                       控制"模式。
10 Mvs P1, -50      '——⑦ Return the robot to the position 50 mm behind P1 in the
                       hand direction using linear interpolation./ 直线插补到辅助
                       工作点，辅助工作点距离 P1 点 50mm。
11 End              '——End the program./ 结束程序。
```

(8) Hand and tool control 抓手控制（图 6-28）

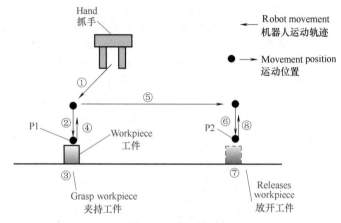

图 6-28　抓手控制

Program explanation 程序说明：

```
1 Tool (0, 0, 95, 0, 0, 0)   '——Set the hand length to 95 mm./ 设置抓手长度为
                                95mm。
2 MOV P1, -50                '——① Move with joint interpolation from P1 to
                                position retracted 50mm in hand direction./
                                关节插补到辅助工作点，辅助工作点距离 P1 点
                                50mm。
3 Ovrd 50                    '——Set the movement speed to half of the maximum
                                speed./ 设置运动速度为最大速度的一半。
4 Mvs P1                     '——② Move with linear interpolation to P1（go
                                to grasp workpiece）./ 直线插补到 P1 点（运动
                                到夹持工件位置）。
5 Dly 0.5                    '——Wait for the 0.5 seconds for the completion of
                                arrival to the target position./ 到达目标位置后，
                                暂停 0.5s。
6 HClose 1                   '——③ Close hand 1（grasp workpiece）./1 号抓手
                                动作（夹持工件）。
7 Dly 0.5                    '——Wait 0.5 seconds./ 等待（暂停）0.5s。
8 Ovrd 100                   '——Set movement speed to maximum speed./ 设
                                置运行速度为最大速度。
```

```
 9 Mvs, -50        '——④ Move linearly from current position（P1）
                      to position retracted 50mm in hand direction
                      （lifts up workpiece）./ 从P1点直线插补运行
                      到辅助工作点，辅助工作点沿抓手方向距离P1点
                      50mm（提起工件）。
10 Mov P2, -50     '——⑤ Move with joint interpolation from P2 to
                      position retracted 50mm in hand direction./
                      关节插补到辅助工作点，辅助工作点距离P2点
                      50mm。
11 Ovrd 50         '——Set movement speed to half of the maximum
                      speed./ 设置运行速度为最大速度的一半。
12 Mvs P2          '——⑥ Move with linear interpolation to P2（go
                      to place workpiece）./ 直线插补到P2点（准备
                      放下工件）。
13 Dly 0.5         '——Wait for the 0.5 seconds for the completion of
                      arrival to the target position./ 暂停0.5s，保证
                      到达目标位置。
14 HOpen 1         '——⑦ Open hand 1（releases workpiece）./ 打开
                      1号抓手（放下工件）。
15 Dly 0.5         '——Wait 0.5 seconds./ 等待0.5s。
16 Ovrd 100        '——Set movement speed to maximum speed./ 设
                      置运行速度为最大速度。
17 MVS, -50        '——⑧ Move linearly from current position（P2）
                      to position retracted 50mm in hand direction
                      （separate from workpiece）./ 从当前位置P2直
                      线插补运动到辅助工作点，辅助工作点沿抓手方向
                      距离P2点50mm（离开工件）。
18 End             '——End the program./ 结束程序。
```

（9）Relative calculation of position data（multiplication）　位置点数据的乘法运算（图6-29）

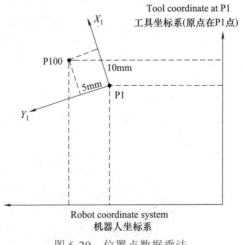

图6-29　位置点数据乘法

Chapter 6　Commands of the Robot

Numerical variables are calculated by the usual four arithmetic operations. The calculation of position variables involves coordinate conversions, however, not just the four basic arithmetic operations. This is explained using simple examples. An example of relative calculation (multiplication) is as follows.

数字变量的计算为四则运算。位置变量的计算包含坐标系变换，不仅仅是四则运算，举例说明如下（乘法）。

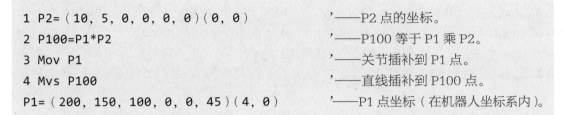

乘法运算的规则：在乘法表达式"P100=P1*P2"中，P1 为机器人坐标系内坐标，P2 点是工具坐标系内的坐标，而 P1 点是工具坐标系的原点。

(10) Relative calculation of position data (addition)　位置点数据的加法运算（图 6-30）

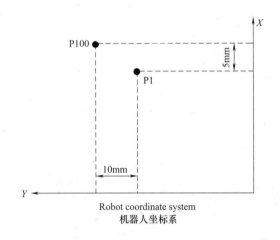

图 6-30　位置点数据加法

Relative calculation of position data (addition).
位置点数据的加法运算。

```
1 P2=(5,10,0,0,0,0)(0,0)           '——P2 点坐标数据。
2 P100=P1+P2                       '——加法运算。
3 Mov P1                           '——关节插补到 P1。
4 Mvs P100                         '——直线插补到 P100。
P1=(200,150,100,0,0,45)(4,0)       '——P1 点坐标。
```

加法运算的规则是：在加法表达式"P100=P1+P2"中，P1 为机器人坐标系内坐标，P2 点也是机器人坐标系内的坐标，而 P100 点的数值是坐标数据的相加。

6.6 Detailed explanation of command words
指令详细解释

（1）Base 基本坐标系指令
Function 功能：
Changes (relocation and rotation) can be made to the world coordinate system which is the basis for the control of the robot's current position.

Base 指令用于以"基本坐标系"为基准建立一个新的"世界坐标系"。Base 指令中的操作数据是在新的世界坐标系中观察到的"基本坐标系"数据，如图 6-31 所示。

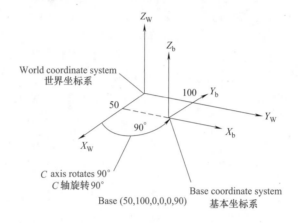

图 6-31 Base 指令示意图 1

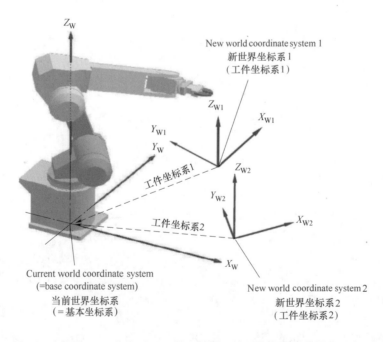

图 6-32 使用 Base 指令以"工件坐标系"编号指定"新的世界坐标系"

Base coordinate number: The system's initial value or value set in the parameter concerned (work coordinate system) is designated as base conversion data.

使用 Base 指令，以"工件坐标系"编号指定"新的世界坐标系"。

例：

1 Base 1

Work coordinate system 1 (parameter: WK1CORD) is defined as a new world coordinate system.

设定"1 号工件坐标系"为"新的世界坐标系"（图 6-32）。

（2）Cmp Pos（Compliance Posture） 柔性控制指令（图 6-33）

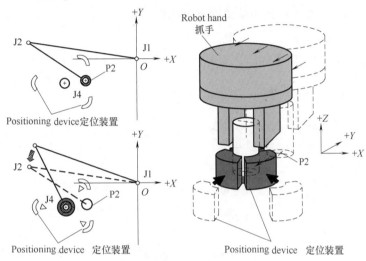

图 6-33 Cmp Pos 指令示意图

Function 功能：

Start the soft control mode (compliance mode) of the specified axis in the XYZ coordinate system.

指定 XYZ 坐标系中的某一轴进入"柔性控制"。

Reference program 参考程序：

1 Mov P1	'——Move in front of the part insertion position./ 移动到 P1 点。
2 CmpG 0.5, 0.5, 1.0, 0.5, 0.5,,,	'——Set softness./ 设置柔性度。
3 Cmp Pos, &B011011	'——The X, Y, A and B axes are put in the state where they are controlled in a pliable manner./ 设置 X、Y、A、B 轴进入柔性控制模式。
4 Mvs P2	'——Move to the part insertion position./ 移动到 P2 点。
5 M_Out, (10) =1	'——Instruct to close the chuck for

	positioning./ 输出（10）=ON。
6 Dly 1.0	'——Wait for the completion of chuck closing（1.0sec）./ 暂停 1.0s。
7 HOpen 1	'——Open the hand. / 张开抓手。
8 Mvs, -100	'——Retreat 100 mm in the Z direction of the tool coordinate system./ 移动到 P2 点上方 100mm。
9 Cmp Off	'——Return to normal state./ 关闭柔性控制模式。

（3）Cmp Tool 工具坐标系中的柔性控制指令（图 6-34）

Function 功能：

Start the soft control mode（compliance mode）of the specified axis in the tool coordinate system.

对工具坐标系中指定的轴启动柔性控制模式。

Cmp Tool, &B011011——指定 X、Y、A、B 轴进入柔性控制模式。

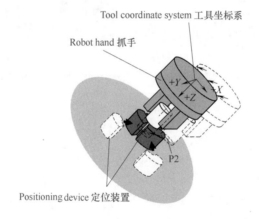

图 6-34　Cmp Tool 指令示意图

（4）Set proximity distance　设置接近距离（图 6-35）

（5）ColChk（Col Check）　碰撞检测指令（图 6-36）

Function 功能：

The collision detection function quickly stops the robot when the robot's hands and/or arms interfere with peripheral devices so as to minimize damage and deformation of the robot's tool part or peripheral devices.

碰撞检测功能：在机器人的抓手或机械臂与周边机器发生干涉时，立即停止机器人，将机器人抓手及周边机器的破损及变形等损坏抑制到最小。

Chapter 6　Commands of the Robot

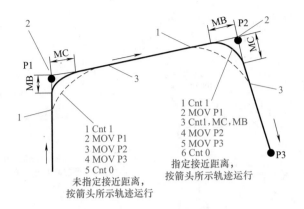

图 6-35　设置接近距离指令示意图

1—Deceleration start position/ 减速起点位置；2—If the MB and MC values are different，connection is made using a value lower than the smaller of these two values/ 当 MB 与 MC 的数值不相同时，取较小值连接；
3—Acceleration end position/ 加速结束位置

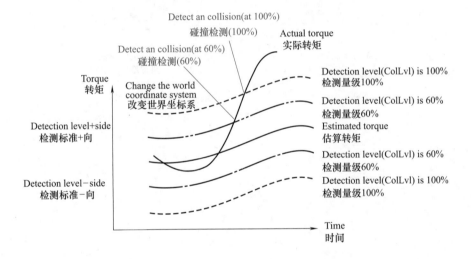

图 6-36　碰撞检测示意图

（6）Stop type　停止类型

1）Stop type 1　停止类型 1（图 6-37）

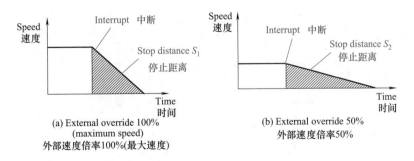

图 6-37　停止类型 1（$S_1=S_2$）

2) Stop type 2　停止类型2（图6-38）

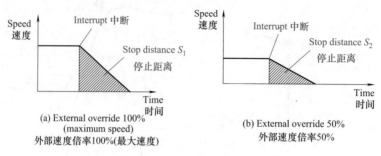

图6-38　停止类型2（$S_1 \neq S_2$）

3) Stop type 3　停止类型3（图6-39）

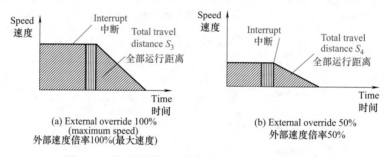

图6-39　停止类型3（执行完成后再停止，$S_3=S_4$）

Chapter 7
Functions Set with Parameters
参数设置

7.1 Standard tool coordinate
标准工具坐标系

(1) Set parameters 设置参数

Tools data must be set if the robot's control point is to be set at the hand tip when the hand is installed on the robot. The setting can be done in the following manners.

1) Set in the MEXTL parameter.

2) Set in the robot program using the tool instruction.

在机器人安装了抓手，机器人的控制点（需要）设置在"抓手前端"时，必须要设定TOOL（抓手）数据（以建立新的"工具坐标系"）。设定方法如下。

1) 以参数 MEXTL 设定。

2) 在机器人程序内以 TOOL 指令设定（图 7-1 是 6 轴机器人的设置样例）。

The default value at the factory default setting is set to zero, where the control point is set to the mechanical interface (flange plane).

出厂值设置为 0，"控制点"为机械 I/F（法兰面）原点。

Structure of tools data: X, Y, Z, A, B, C.

X, Y and Z axis: Shift from the mechanical interface in the tool coordinate system.

A axis: X-axis rotation in the tool coordinate system.

B axis: Y-axis rotation in the tool coordinate system.

C axis: Z-axis rotation in the tool coordinate system.

TOOL 数据的构成：X，Y，Z，A，B，C。

X，Y，Z 轴：以机械 I/F 坐标系为基准的"抓手前端（控制点）"坐标值。

A 轴：绕 X 轴旋转的角度。

B 轴：绕 Y 轴旋转的角度。

C 轴：绕 Z 轴旋转的角度。

（2）A case for a vertical 6-axis robot　6 轴机器人设置样例（图 7-1）。

1）Sample parameter setting：

Parameter name：MEXTL．

Value：0，0，95，0，0，0．

2）Sample Tool instruction setting：

$$1\,\text{Tool}\,(0,\,0,\,95,\,0,\,0,\,0)$$

6-axis robot can take various postures within the movement range.

1）参数的设置样例：

参数名：MEXTL。

参数值：0，0，95，0，0，0。

2）Tool 指令设置样例：

$$1\,\text{Tool}\,(0,\,0,\,95,\,0,\,0,\,0)$$

6 轴机器人在移动范围内可作出各种姿势。

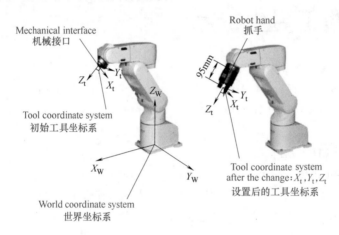

图 7-1　6 轴机器人工具坐标系设置示意图

（3）A case for a vertical 5-axis robot　5 轴机器人设置样例（图 7-2）

1）Sample parameter setting：

Parameter name：MEXTL．

Value：0，0，95，0，0，0．

2）Sample Tool instruction setting：

$$1\,\text{Tool}\,(0,\,0,\,95,\,0,\,0,\,0)$$

Only the Z axis component is valid for a 5-axis robot for movement range reasons. Data input to other axes will be ignored.

1）参数的设置样例：

参数名：MEXTL。

参数值：0，0，95，0，0，0。

2）Tool 指令的设置样例：

$$1\,\text{Tool}\,(0,\,0,\,95,\,0,\,0,\,0)$$

5 轴机器人因动作范围关系，只对 Z 轴成分有效，其他轴输入数据无效。

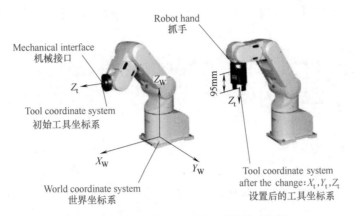

图 7-2　5 轴机器人工具坐标系设置示意图

（4）A case for a horizontal 4-axis robot　水平 4 轴机器人设置样例（图 7-3）

1) Sample parameter setting:

Parameter name: MEXTL.

Value: 0, 0, -10, 0, 0, 0.

2) Sample Tool instruction setting:

$$1\ \text{Tool}\ (0,\ 0,\ -10,\ 0,\ 0,\ 0)$$

Horizontal 4-axis robots can basically offset using parallel shifting. Note that the orientation of the tool coordinate system is set up differently from that of vertical robots.

1) 参数设置样例：

参数名：MEXTL。

参数值：0, 0, -10, 0, 0, 0。

2) TOOL 指令设置样例：

$$1\ \text{Tool}\ (0,\ 0,\ -10,\ 0,\ 0,\ 0)$$

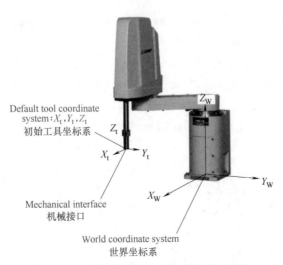

图 7-3　4 轴机器人工具坐标系设置示意图

水平4轴机器人基本上可以使用与"工作轴"运动方向平行的OFFSET设置。水平4轴机器人的设置与垂直机器人的TOOL坐标方向不同，必须特别注意。

7.2 Standard base coordinate 标准基本坐标系

（1）Definition 定义

The position of the world coordinate system is set to zero (0) before leaving the factory, and therefore, the base coordinate system (robot's installation position) is in agreement with the world coordinate system (coordinate system which is the basis for robot's current position). By utilizing the base conversion function, you can set the origin point of the world coordinate system at a location other than the center of the J1 axis.

机器人出厂时"世界坐标系"位置设置为0，基本坐标系（机器人的安装位置）与世界坐标系一致。使用基本转换功能，可将"世界坐标系原点"设定在J1轴的中心位置以外。

Four methods are available for setting the world coordinate system:
1) Specifying parameter MEXBS directly with base conversion data.
2) Specifying parameter MEXBSNO with a base coordinate number.
3) Specifying the J1 axis offset angle using parameter J1OFFSET (vertical 5-axis type robot only).
4) Executing a relevant base command under the robot program.

设定世界坐标系的方法有以下4种：
1) 使用参数：MEXBS 直接设定"基本转换数据"。
2) 使用参数：MEXBSNO 设置坐标系编号。
3) 使用参数：J1OFFSET 设定J1轴偏置角度（仅限垂直5轴型的机器人）。
4) 在机器人程序中执行 Base 指令进行设定。

Structure of base coordinate system data: X, Y, Z, A, B and C.

X, Y and Z axis: The position of robot coordinate system from the base coordinate system origin.

A axis: X axis rotation in the world coordinate system.

B axis: Y axis rotation in the world coordinate system.

C axis: Z axis rotation in the world coordinate system.

"基本坐标系"数据的构成：X，Y，Z，A，B，C。

X，Y，Z 轴：从世界坐标系观察到的"基本坐标系"位置。

A 轴：绕世界坐标系 X 轴的旋转角度。

B 轴：绕世界坐标系 X 轴的旋转角度。

C 轴：绕世界坐标系 X 轴的旋转角度。

(2) Example 设置样例（图7-4）
1) Sample parameter setting:
Parameter name: MEXBS.
Value: 100, 150, 0, 0, 0, -30.
2) Sample Base instruction setting:
$$1\ Base\ (100,\ 150,\ 0,\ 0,\ 0,\ -30)$$
1) 使用参数设置：
参数名：MEXBS。
参数值：100, 150, 0, 0, 0, -30。
2) 使用 Base 指令设置：
$$1\ Base\ (100,\ 150,\ 0,\ 0,\ 0,\ -30)$$

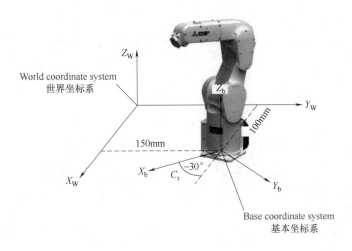

图 7-4 世界坐标系的设置

(3) Setting base coordinate system 设置"基本坐标系"

This choice is accomplished by specifying by the parameter AREAnCS that the reference coordinate system is "world coordinate system" (for moving user-defined area concurrently) or that the same is "base coordinate system" (keeping user-defined area fixed).

使用参数：AREAnCS 可设置"世界坐标系"或"基本坐标系"作为参考坐标系。

Change to base coordinate system causes a change to relative positional relation between the robot body and the user-defined area.

选择"世界坐标系"作为参考坐标系时，改变世界坐标系会引起"用户定义区"与"机器人"相对位置的改变，如图 7-5 所示。

(4) Changing base coordinate system 改变基本坐标系

Changing base coordinate system causes no change to relative positional relation between the robot body and the user-defined area.

改变基本坐标系不会引起"用户定义区"与"机器人"相对位置的改变，如图 7-6 所示。

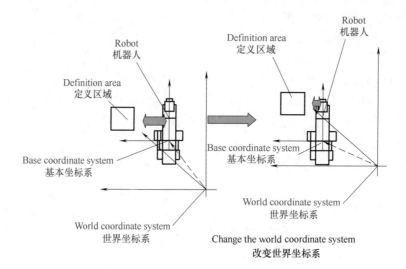

图 7-5 选择世界坐标系作参考坐标系引起的改变

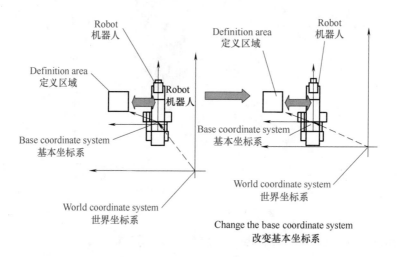

图 7-6 改变基准坐标系不会引起"用户定义区"与"机器人"相对位置的改变

(5) Setting Areas 设置区域

Areas to be set include position area, posture area and additional axis area.
设置区域分为位置区域、"立体形位"区域和附加轴区域。

The following is a description of the steps that are followed to set these areas.
以下是设置部分区域的步骤。

1) Position area 位置区域

A position area for user-defined area is defined by the coordinate of a diagonal point which is determined by the elements X, Y and Z in the parameters AREA*P1 and AREA*P2 (* is 1 to 32).

The coordinate values thus determined are those which refer to the coordinate system selected by the parameter AREA*CS (* is 1 to 32).

使用参数 AREA*CS (*=1~32) 选择坐标系。

Chapter 7 Functions Set with Parameters

使用参数 AREA*P1 和 AREA*P2（*=1~32）的 X、Y、Z 设置对角点坐标，从而设置用户区域，如图 7-7 所示。

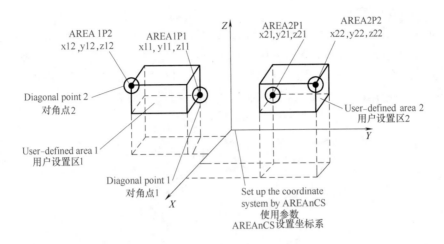

图 7-7 用参数设置用户区域

2) Posture Area   形位区域

A posture area for the user-defined area is defined by specifying elements A, B and C in the parameters AREAnP1 and AREAnP2. Set up the value based on the coordinate system selected by AREAnCS.

使用参数 AREAnCS 选择坐标系，使用参数 AREAnP1 和 AREAnP2 设置旋转角度 A、B、C 的旋转范围数值，见图 7-8。

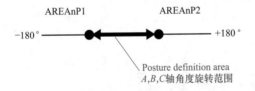

(a) If the relative locations of posture elements are set for
AREAnP2>AREAnP1
如果旋转角度的关系设置为"AREAnP2>AREAnP1"

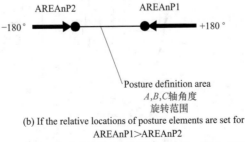

(b) If the relative locations of posture elements are set for
AREAnP1>AREAnP2
如果旋转角度的关系设置为"AREAnP1>AREAnP2"

图 7-8 用"立体形位"区域设置坐标系

7.3 Free plane limit　自由平面限制

(1) Definition　定义

Define any plane in the world coordinate system, determine the front or back of the plane, and generate a free plane limit error.

在世界坐标系内定义一个平面，检测（控制点运动范围）是在这个平面的前面还是后面并发出报警，这种方法称为"自由平面限制"。"自由平面"的设置方法见图7-9。

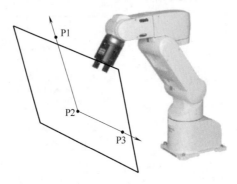

图7-9　自由平面限制

(2) Selection of a coordinate system for a free plane limit　自由平面限制的坐标系选择

1) Free plan limit after changing the base coordinate system.

改换基本坐标系后的自由平面限制（图7-10）。

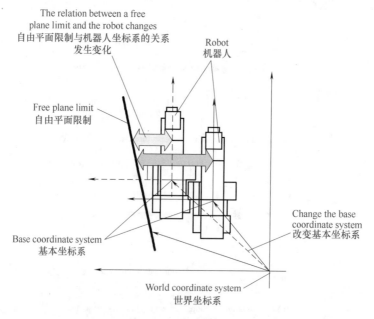

图7-10　改换基本坐标系后的自由平面限制

2) Free plane limit after changing the world coordinate system.

改换世界坐标系后的自由平面限制区域（图7-11）。

Chapter 7　Functions Set with Parameters

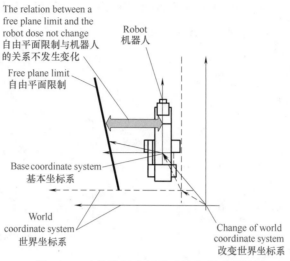

图 7-11　改换世界坐标系后的自由平面限制

7.4　Automatic return setting after jog feed at pause
点动暂停后返回轨迹设置

（1）Outline　概述

Automatic return setting after jog feed at pause:

This specifies the path behavior that takes place when the robot is paused during automatic operation or during step feed operation, moved to a different position using a jog feed with T/B, and the automatic operation is resumed or the step feed operation is executed again. See the following diagram.

设置点动运行停止后重新启动的运行轨迹：

机器人在自动运行或单步进给时发生暂停，使用示教单元向某一位置点动进给，在到达该位置后再重新自动运行，或重新执行单步进给，可使用参数 RETPATH 设置重新启动后的运行轨迹。运行轨迹如图 7-12 所示。

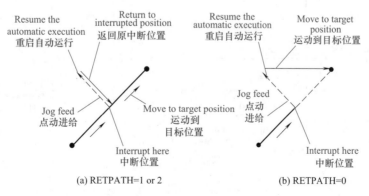

图 7-12　设置点动运行停止后重新启动的运行轨迹

（2）Automatic return setting after jog feed at pause　点动暂停后返回轨迹设置（图 7-13）

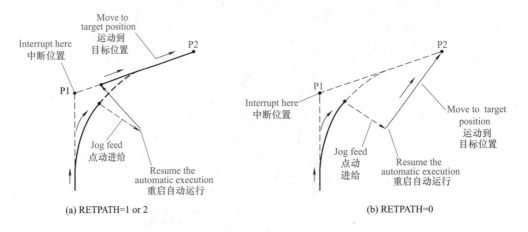

图 7-13　点动暂停后返回轨迹设置

7.5　Warm-up operation mode 暖机运行模式

The warm-up operation mode is the function that operates the robot at a reduced speed immediately after powering on the controller and gradually returns to the original speed as the operation time elapses.

暖机运行模式：控制器的电源 ON 后立即降低速度运行，经过一段时间后再恢复到原来的速度，如图 7-14 所示。在低温下运行及长期停止后再运行时，如果发生"误差过大报警"时，需执行暖机运行模式。

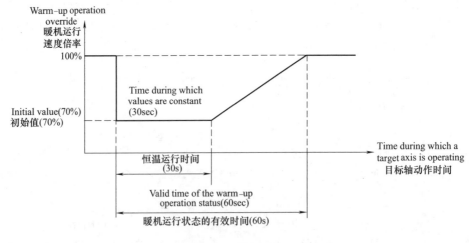

图 7-14　暖机运行

7.6 High-speed RAM operation image 高速 RAM 运行图（图 7-15）

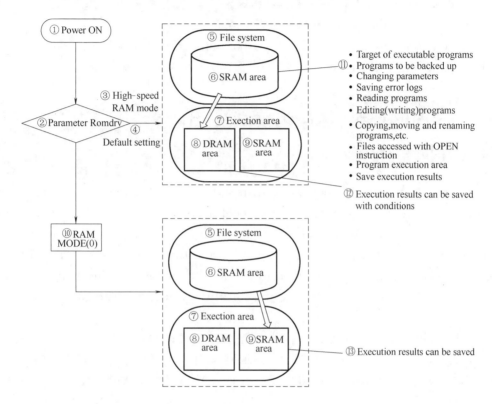

图 7-15　高速 RAM 运行图

① Power ON——电源 =ON。

② Parameter Romdrv——模式选择参数。Romdrv 用于选择"High-speed RAM mode"或"RAM mode"。

③ High-speed RAM mode——高速 RAM 模式。

④ Default setting——初始设置。

⑤ File system——文件系统。

⑥ SRAM area——SRAM 区。

⑦ Exection area——执行区。

⑧ DRAM area——DRAM 区。

⑨ SRAM area——SRAM 区。

⑩ RAM MODE（0）——RAM 模式。

⑪• Target of executable programs——可执行程序的对象。

• Programs to be backed up——备份程序。

• Changing parameters——修改参数。

• Saving error logs——保存报警日志。

- Reading programs——读取程序。
- Editing (writing) programs——编辑程序。
- Copying, moving and renaming programs, etc.——复制、移动、重命名程序。
- Files accessed with OPEN instruction——使用 OPEN 指令存取文件。
- Program execution area——程序执行区。
- Save execution results——保存执行结果。

⑫ Execution results can be saved with conditions——根据条件保存执行结果。

⑬ Execution results can be saved——保存执行结果。

7.7 Collision detection function 碰撞检测

The collision detection function detects interferences using such servo characteristics. First, the torque required for each joint axis is estimated based on the current position instruction and load setting. Next, the values are compared with the actually generated torques for each axis one by one. If the difference exceeds the allowable range (detection level), the function judges that an interference occurred. It immediately turns the servo off and stops the robot.

碰撞检测功能是使用伺服系统的特性检测干涉。首先，依据当前位置指令和各轴负载，计算出各关节轴所需要的转矩；其次，将各轴计算转矩与实际转矩依次比较。如果有某一轴的差值超过容许范围（检测量级），则判断发生干涉，机器人立即关闭伺服系统并且紧急停止（图 7-16）。

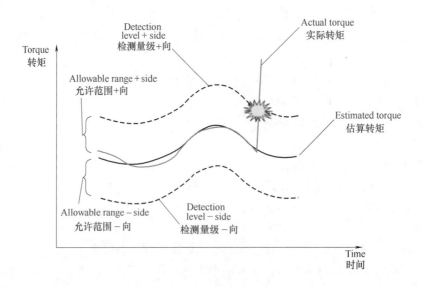

图 7-16　碰撞检测示意图

7.8 Simulation and robot calibration operation
模拟及机器人校准运行

（1）Connect of equipment　设备连接（图7-17）

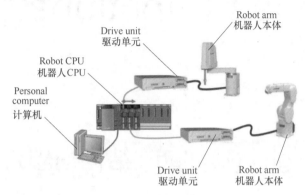

图7-17　设备连接示意图

（2）Simulation component　模拟构件（图7-18）

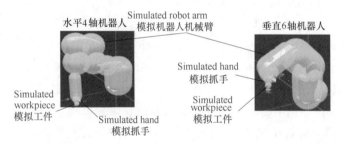

图7-18　模拟构件意图

（3）Calibration between robots　机器人之间的校准（图7-19）

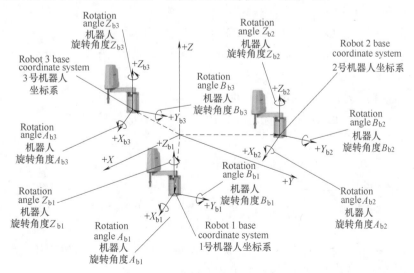

图7-19　机器人之间的校准

Chapter 8
Advanced Functions
高级功能

8.1 Configuration flag 结构标志

8.1.1 Configuration flag of 6-axis robot 6 轴机器人的结构标志

The configuration flag indicates the robot posture. For the 6-axis robot, the robot hand end is saved with the position data configured of X, Y, Z, A, B and C. However, even with the same position data, there are several postures that the robot can change to. The posture is expressed by this configuration flag, and the posture is saved with FL1 in the position constant (X, Y, Z, A, B, C), (FL1, FL2). The types of configuration flags are shown below.

结构标志用于表示机器人的"立体形位"。机器人的抓手工作点的位置使用基于 X、Y、Z、A、B、C 的位置数据进行表示。但是,即使相同的位置数据,机器人可采取的"立体形位"也有多个。表示"立体形位"的标志即"结构标志",通过位置数据(X,Y,Z,A,B,C)、(FL1,FL2)中的"FL1"表示"结构标志"。

(1) Right/Left 右/左

P is center of J5 axis rotation in comparison with the plane through the J1 axis vertical to the ground.

表示 J5 轴旋转中心(P)相对于 J1 轴的旋转中心的位置(图 8-1)。

FL1(Flag 1)&B0000 0000 标志的 bit2 表示"左右标志":

Bit2=1——Right;

Bit2=0——Left。

(2) Above/Below 上/下

P is center of J5 axis rotation in comparison with the plane through both the J3 and the J2 axis.

"上下标志"表示 J5 轴旋转中心(P)相对于 J2 轴和 J3 轴构成平面的相对位置(图 8-2)。

FL1(Flag 1)&B0000 0000 标志的 bit1 表示"上下标志":

Bit1=1——Above;
Bit1=0——Below。

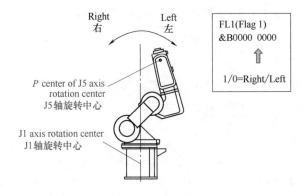

图 8-1 左右标志

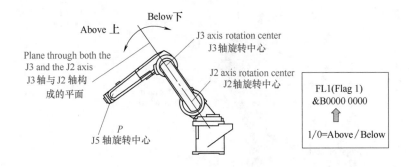

图 8-2 上下标志

（3）Nonflip/Flip 法兰面上下（图 8-3）

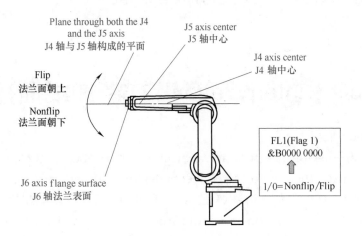

图 8-3 法兰面朝向标志

This means in which side the J6 axis is in comparison with the plane through both the J4 and the J5 axis.

Nonflip/Flip 表示 J6 轴法兰面相对于通过 J4 轴和 J5 轴中心线构成的平面的相对位置。

FL1（Flag 1）&B0000 0000 标志的 bit0 表示"NONFLIP/FLIP 标志"：

Bit0=1——Nonflip；

Bit0=0——Flip。

8.1.2　Configuration flag for horizontal multi-joint robot　水平多关节型机器人的结构标志

Right/Left 右 / 左：

Indicates the location of the end axis relative to the line that passes through both the rotational center of the J1 axis and the rotation center of the J2 axis.

以 J1 轴旋转中心和 J2 轴旋转中心的连接线为基准，以"右 / 左"表示工作轴相对于"连接线"的位置（图 8-4）。

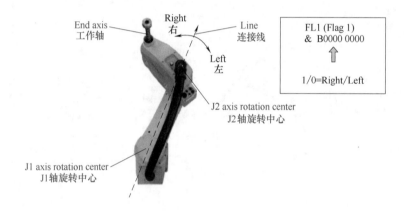

图 8-4　左右标志

FL1（Flag 1）&B0000 0000 标志的 bit2 表示"左右标志"：

Bit2=1——Right；

Bit2=0——Left。

8.2　Spline interpolation　样条曲线插补

(1) Outline　概要

Spline interpolation is a function that moves the robot at the designated speed along a spline curve that smoothly connects designated path points.

样条曲线插补，即机器人沿设定的轨迹点构成的平滑曲线，以设置的速度运动的功能。

Ex-T spline interpolation is a function used to move the spline curve that smoothly links path points specified on the workpiece grasped by the robot along an arbitrary coordinate system origin（Ex-T coordinate system origin）at a specified speed (Ex-T coordinate system origin viewed from workpiece moves relatively at specified speed).

Ex-T 样条曲线插补，即机器人夹持工件在 Ex-T 坐标系内以 Ex-T 坐标系原点为基准做平滑曲线插补。平滑曲线的轨迹由工件确定。

The robot can be moved along a curved path that was not possible with conventional linear or circular interpolation. This interpolation can be used for sealing, polishing and chamfering processes, etc.

机器人可以沿着曲线轨迹运行，这种曲线插补是常规的直线插补和圆弧插补无法完成的。样条曲线插补常用于密封、抛光、倒角等作业。

(2) Sweep of spling interpolation　样条插补的轮廓线（图 8-5、图 8-6）

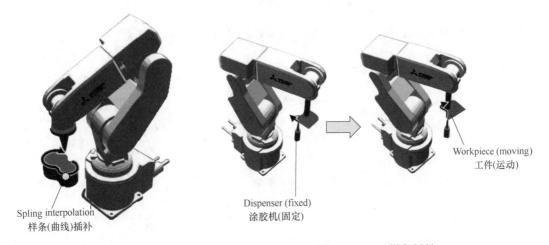

图 8-5　样条插补的轮廓线　　　　　图 8-6　Ex-T 样条插补

(3) Features　特点

A smooth spline curve is generated between each path point so that the robot positions and posture designated as path points are passed. The robot moves along that curve at the speed designated with linear speed.

由各轨迹点生成一条平滑的样条曲线，这样可以确定机器人的位置点和"立体形位"，机器人即以设定的速度沿曲线插补，如图 8-7 所示。

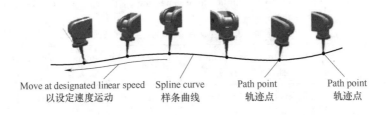

图 8-7　样条插补特点

（4）Generate a program file for the spline interpolation 为插补轨迹生成一个程序文件（图8-8）

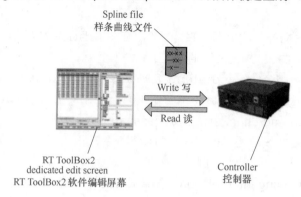

图8-8 为插补轨迹生成一个程序文件

（5）Spline interpolation terminology 样条插补术语（图8-9、表8-1）

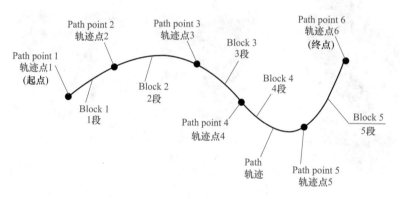

图8-9 样条插补术语

表8-1 样条插补术语

Terminology 术语	Explanation 说明
Path point 轨迹点	This is the robot position data（Cartesian coordinate value）used to generate the spline curve 为生成样条曲线而使用的机器人的位置数据（直角坐标值）
Block 段	Refers to the curve line or segment generated between two adjacent path points 相邻2个轨迹点之间所生成的曲线或线段
Path 轨迹	Refers to the spline curve generated by the spline interpolation command and passes through path points 通过样条插补指令生成的通过轨迹点的样条曲线
Start position 起点	Refers to the path point where spline interpolation movement starts 样条插补动作的启始轨迹点
End position 终点	Refers to the path point where spline interpolation movement ends 样条插补动作的终点
Spline file 样条曲线文件	File containing the path points and setting values required for execution，etc. One file corresponds to one spline interpolation 存有轨迹点及执行时所需要的设定值等的文件，1个文件对应1个样条插补

续表

Terminology 术语	Explanation 说明
Spline file edit screen 样条文件编辑画面	RT ToolBox2 screen dedicated for creating, editing and saving spline file RT ToolBox2 中配备的用于创建、编辑、保存样条文件的专用画面
DXF file import screen DXF 文件导入画面	RT ToolBox2 screen dedicated for creating, editing and saving spline file from a DXF file RT ToolBox2 中配备的用于从 DXF 文件创建、保存样条文件的专用画面
Ex-T coordinate system origin Ex-T 坐标系原点	This is the coordinate system origin that is the subject of arbitrarily defined control outside the robot Ex-T 坐标系原点

（6）The action of robot in spline interpolation　样条插补中机器人的动作（图 8-10）

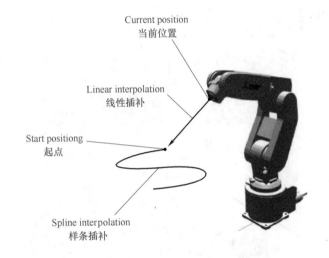

图 8-10　样条插补中机器人的动作

（7）Halt operation and resume operation　中断和重新启动（图 8-11）

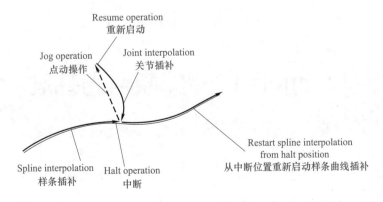

图 8-11　中断和重新启动时的动作

（8）Signal output　信号输出（图 8-12）

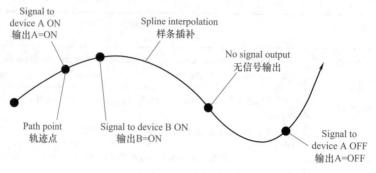

图 8-12　信号输出

（9）Graphical spline interpolation work details　图解样条插补工作细节（图 8-13）

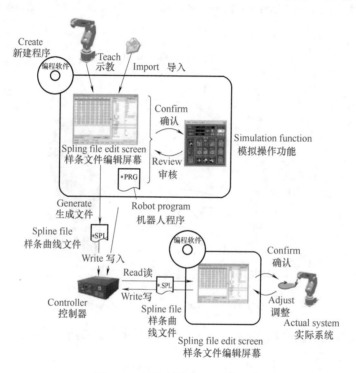

图 8-13　图解样条插补工作细节

8.3　Ex-T control　以外部原点为基准的控制

8.3.1　Outline　概述

The Ex-T control is the function to operate the robot using the origin of the externally fixed coordinate system as the robot control point. The examples of applications include the following.

Ex-T 控制功能是以固定于外部的坐标系的原点为机器人的控制点，使机器人动作，用于以下作业。

Polishing　The robot holds the target workpiece, and pushes it against the fixed grinder or abrasive belt for deburring or surface finishing.

研磨作业　机器人夹持着工件，使工件触压到固定安装的砂轮或砂带上进行去毛刺或表面精加工。

Coating　The robot holds the target workpiece, and applies solvent or adhesive supplied from the fixed dispenser to the workpiece.

涂敷作业　机器人夹持着工件，通过固定安装的点胶机对工件涂敷溶剂或黏合剂的作业。

（1）Polishing　抛光（图 8-14）

（2）Coating　涂胶（图 8-15）

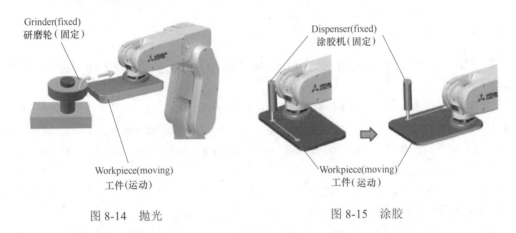

图 8-14　抛光　　　　　　　　图 8-15　涂胶

8.3.2　Ex-T coordinate setting　Ex-T 坐标系设置

（1）Externally fixed coordinate system setting　外部固定坐标系设置（图 8-16）

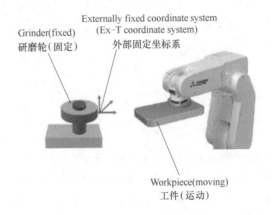

图 8-16　外部固定坐标系设置

（2）Externally fixed coordinate system　外部固定坐标系（图 8-17）

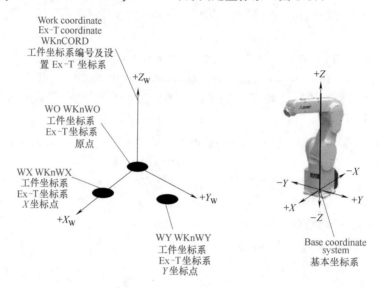

图 8-17　外部固定坐标系

（3）Movement of the posture element in the WORK jog　在工件 jog 中"立体形位"的运动

The jog operation of the posture element in the WORK jog is rotation around the axes parallel to the X, Y, and Z axes of the work coordinate at the control point. The position remains fixed. Fig. 8-18 shows the example of C element movement in the WORK jog.

工件 JOG 中的"立体形位"部分的 JOG 动作，是控制点绕工件坐标系的 XYZ 轴的平行轴的旋转。此时，位置固定不变。工件 JOG 中 C 轴动作示例如图 8-18 所示。

WO-WX-WY 表示工件坐标系（从 +WZ 观察到的工件坐标系俯视图）。TCP 为机器人的控制点，四边形为机器人夹持的工件，虚线表示移动后的工件位置（"立体形位"）。

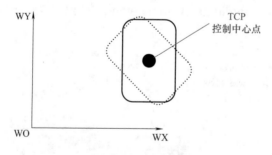

图 8-18　在工件坐标系中绕 C 轴运动

（4）Movement of the posture element in the Ex-T jog　在 Ex-T 点动中的运动"形位"

The jog operation of the posture element in the Ex-T jog is rotation around the X, Y, and Z axes of the Ex-T coordinate system (work coordinate system). The robot position also changes. Fig. 8-19 shows the example of C element movement in the Ex-T jog.

Ex-T jog 中的"立体形位"部分的 JOG 动作，为绕 Ex-T 坐标系（工件坐标系）的 XYZ 轴的旋转。此时，机器人的位置也发生变化。

Ex-T jog 中 C 轴旋转动作示例见图 8-19。

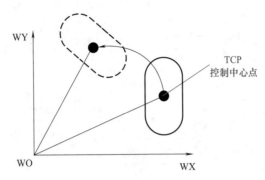

图 8-19　在 Ex-T 坐标系中的 C 轴旋转运动

（5）Rotating operation in Ex-T coordinate system　在 Ex-T 坐标系中的 C 轴旋转运动

In Fig. 8-20, the origin of the Ex-T coordinate system (work coordinate system) is located on the workpiece. Rotating operation is performed around WO in both cases.

在图 8-20 中，Ex-T 坐标系（工件坐标系）的原点（或 Z 轴）与工件接触，以 WO 为中心进行旋转动作。

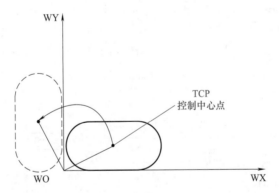

图 8-20　在 Ex-T 坐标系中的 C 轴旋转运动

8.3.3　Operation　操作

8.3.3.1　WORK jog operation of the 6-axis robot　6 轴机器人在工件坐标系中的点动

（1）WORK jog operation of the 6-axis robot　6 轴机器人在工件坐标系中的点动

When the X, Y, or Z keys are used, the operation is the same in the WORK jog and the Ex-T jog modes.

如果只做 XYZ 方向的运动，WORK jog 与 Ex-T jog 操作相同（图 8-21）。

（2）Change the direction of the flange　改变法兰方位

When the A, B, and C keys are used, the operation is different in the WORK jog and the Ex-T jog modes.

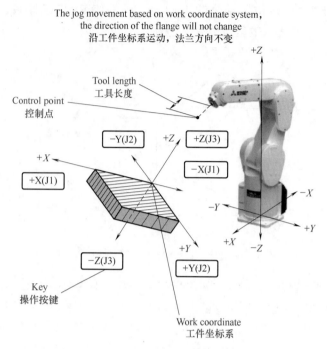

图 8-21　6 轴机器人在工件坐标系中的运动

如果做 ABC 轴的旋转运动，"WORK jog"与"Ex-T jog"操作不相同（图 8-22）。

In WORK jog, the position of the control point does not change.Change the direction of the flange in accordance with the work coordinate system.

在 WORK jog 中，控制点的位置不变，只改变法兰面的方向（图 8-22）。

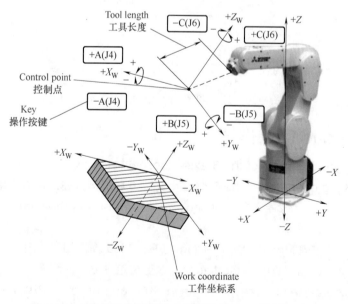

图 8-22　6 轴机器人在工件坐标系中绕 ABC 轴的运动

(3) *A* axis rotates of the control point rotates　在 Ex-T 坐标系的 *A* 向旋转运动（图 8-23）

The control point rotates around each axis of work coordinate system (Ex-T coordinates system).

控制点绕 Ex-T 坐标系的坐标轴旋转。

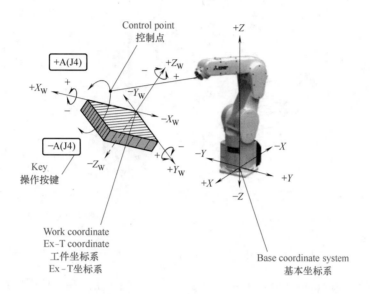

图 8-23　6 轴机器人在 Ex-T 坐标系的 *A* 向旋转运动

(4) *B* axis rotates of the control point rotates　在 Ex-T 坐标系的 *B* 向旋转运动（图 8-24）

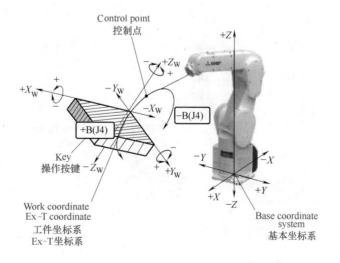

图 8-24　6 轴机器人在 Ex-T 坐标系的 *B* 向旋转运动

（5）C axis rotates of the control point rotates　在Ex-T坐标系的C向旋转运动（图8-25）

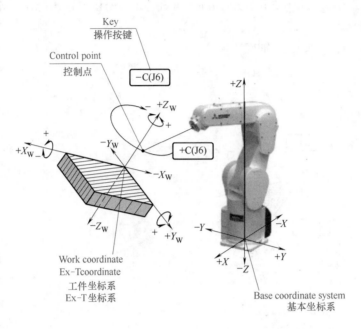

图8-25　6轴机器人在EX-T坐标系的C向旋转运动

8.3.3.2　WORK jog operation of the 4-axis robot　4轴机器人在工件坐标系中的点动

（1）Ex-T jog operation of the 4-axis robot　4轴机器人在Ex-T坐标系的运动

When the XYZ keys are used, the operation is the same in the WORK jog and the Ex-T jog modes.

如果只做XYZ方向的运动，WORK jog 与 Ex-T jog 方式相同（图8-26）。

The jog movement based on work coordinate system.

JOG动作都以"工件坐标系"为基准动作。

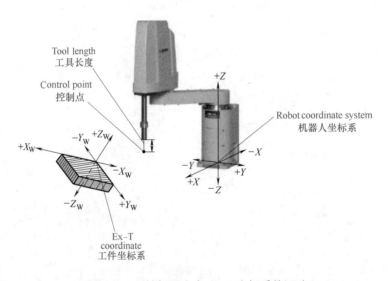

图8-26　4轴机器人在Ex-T坐标系的运动

(2) Movement of the posture element in the Ex-T jog　在工件坐标系（Ex-T 坐标系）中点动（改变"立体形位"）(图 8-27)

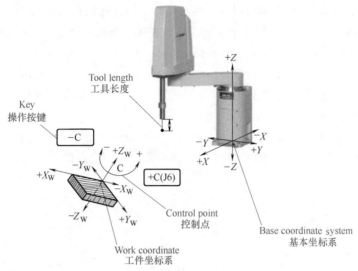

图 8-27　在工件坐标系（Ex-T 坐标系）中点动（改变"立体形位"）

When the [+C (J6)] key is pressed, the control point will rotate in the plus direction around the Z axis (Z_W) of work coordinate system (Ex-T coordinate system). When the [-C (J6)] key is pressed, the control point will rotate in the minus direction.

按下 [+C（J6）] 键，控制点绕工件坐标系（Ex-T 坐标系）Z 轴正向旋转。按下 [-C（J6）] 键，控制点绕工件坐标系（Ex-T 坐标系）Z 轴负向旋转。

(3) WORK jog operation of the 4-axis hanging type　悬挂型 4 轴机器人的"WORK jog"动作

When the XYZ keys are used, the operation is the same in the WORK jog and the Ex-T jog modes.

如果只做 XYZ 方向的运动，"WORK jog"与"Ex-T jog"操作方式相同（图 8-28）。

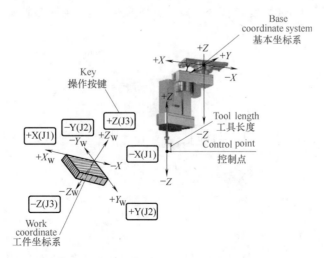

图 8-28　悬挂型 4 轴机器人以工件坐标系为基准的 XYZ 运动

The direction of the end axis will not change.Move the control point with a straight line in accordance with the work coordinate system.

工作轴的运动方向不变，控制点以"工件坐标系"为基准做直线运动。

（4）The action of C instruction in work jog　在工件坐标系"点动"中C指令的动作

When the C key is used, the operation is different in the WORK jog and the Ex-T jog modes. The robot does not move when the A or B key is used.

如果发出C向旋转指令，"工件坐标系点动"与"Ex-T jog"的动作方式是不同的。机器人不响应A/B轴的指令。

The position of the control point does not change.The end axis is rotated.

在"工件坐标系点动"中，如果发出C向旋转指令，控制点位置不变，只是工作轴旋转（图8-29）。

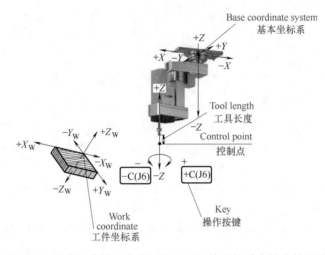

图8-29　改变工作轴"立体形位"以工件坐标系为基准的运动

（5）C-axis Motion based on Ex-T workpiece coordinate system　以Ex-T工件坐标系为基准的C轴运动

在"Ex-T jog"中，如果发C指令，控制点绕Ex-T坐标系的Z轴旋转（图8-30）。

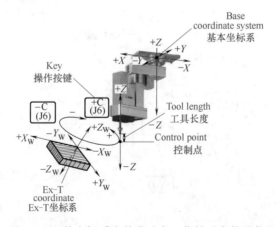

图8-30　以Ex-T工件坐标系为基准改变工作轴"立体形位"的运动

8.3.3.3 Example 加工样例

The example shows a program to perform the operation as shown in the figure. The robot holds the workpiece, and moves the workpiece along the fixed processing tool.

本样例表示一个加工过程。机器人夹持工件沿固定的抛光轮运动。

（1）Example 1　样例 1（图 8-31）

（2）Example 2　样例 2（图 8-32）

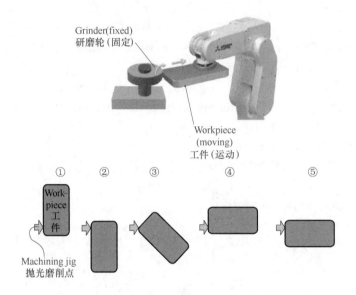

图 8-31　样例 1

8.3.3.4 Setting 设置

（1）Setting 设置

Step 1: Setting of the work coordinate (Ex-T coordinate).

Specify the work coordinate (Ex-T coordinate) so that the contact point between the processing tool and the workpiece shown in the figure is used as the origin of the work coordinate (Ex-T coordinate) (on the work coordinate 1 in this example).

In order to enable jog operation along this work coordinate system, set "1 (Ex-T jog mode)" in the parameter WK1JOGMD.

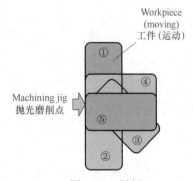

图 8-32　样例 2

步骤 1：设置工件坐标系（Ex-T 坐标系）。

设置工件坐标系（Ex-T 坐标系），设置图中抛光轮与工件接触的部位为工件坐标系（Ex-T 坐标系）原点（将此坐标系设定为工件坐标系 1）。见图 8-33。

为了可以进行基于该工件坐标系的 JOG 动作，必须设置参数 WK1JOGMD =1 [1 (Ex-T JOG 模式)]。

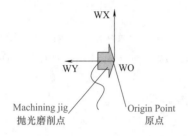

图 8-33 设置抛光轮磨削点为原点

Step 2: Teaching of the positions.

Let the robot actually hold the workpiece for teaching of the positions. The positions ① to ⑤ are taught in this example. For teaching of the positions, performing jog feed of the robot in WORK jog (Ex-T jog) enables jog feed along the processing tool.

步骤 2：位置的示教。

让机器人实际夹持工件，进行位置的示教。本样例中，对图 8-32 的①~⑤的位置进行示教。进行位置示教时，如执行工件 JOG（Ex-T jog），则可以进行基于抛光轮的 JOG 进给。

Step3: Creation of the program.

Create the MELFA-BASIC V program. Some of the actual necessary operations such as workpiece holding movement and input/output of signals are omitted in this example.

步骤 3：编制程序。

编制 MELFA-BASIC V 程序。实际上，还需要工件的夹持动作及信号输入输出等，但此处省略了。

（2）Example program　样例程序

Mov P001	'——Moves to the position ①./ 运动到 P001 点。
Dly 0.5	'——暂停 0.5s。
Spd 50	'——Sets the processing speed (workpiece moving speed) to 50 mm/s./ 设置工件运行速度为 50mm/s。
EMvs 1, P002	'——Moves to the target position ② along the work coordinate 1 by Ex-T linear interpolation./ 沿工件坐标系 1 按 Ex-T 线性插补运行到目标位置"P002"
EMvr 1, P002, P003, P004	'——Moves to the target position ④ from position ② via position ③ along the work coordinate 1 by Ex-T circular arc interpolation./ 沿工件坐标系 1，根据 P002, P003, P004 点按 Ex-T 圆弧插补运行到目标位置"P004"。
EMvs 1, P005	'——Moves to the target position ⑤ along the work coordinate 1 by Ex-T linear interpolation./ 沿工件坐标系 1 按 Ex-T 线性插补运行到目标位置"P005"
...	

In this example, the teaching positions are five.

在本样例中，示教位置点有 5 点。

8.4 Cooperative operation function
联合操作功能

(1) Outline 概述

The cooperative operation function by two robots enables the transportation that two robots grasp the target workpiece at one end, respectively, together in synchronization.

由 2 台机器人执行的"联合操作"能够传送由 2 台机器人夹持的工件同步进行动作。

A position-tracking control of robots enables this operation.

机器人的位置跟踪控制能够执行这种操作。

After the common coordinate are set in a master robot (robot No.1) and a slave robot (robot No.2), robot No.2 obtains the current position data of robot No.1 every controller control time (approximately 7.1ms) via a PLC, and tracks robot No.1 operation.

在对主站机器人(No.1 机器人)和从站机器人(No.2 机器人)设置了共同的坐标后，No.2 机器人经过 PLC 获取 No.1 机器人的"当前位置数据"，获取周期为约 7.1ms，这样跟踪 No.1 机器人的动作。

(2) Cooperative operation 联合操作（图 8-34）

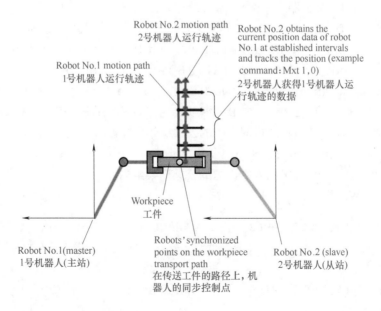

图 8-34 联合操作

（3）System configuration　系统构成（图 8-35）

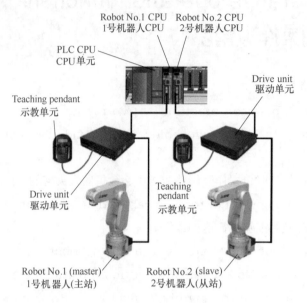

图 8-35　系统构成

（4）Adjustment 1　调整 1（图 8-36）

Adjustment of the common base coordinate:

Use the robot program BFRM.prg to set the common coordinate in robot No.1 and robot No.2.Use the setting of the robot No.1 base coordinate for robot No.2 base coordinates.

调整公共基本坐标：使用机器人程序 BFRM.prg 在 No.1 机器人和 No.2 机器人中设置公共基本坐标。对于 No.2 机器人基本坐标，设置 No.1 机器的基本坐标。

Outline of setting procedure　设置方法概述：

1）Set the position data（PPL1, PPL2 and PPL3）to define the common frame coordinate in robot No.1 and robot No.2.Specify the common position data in the common frame coordinate of robot No.1 and robot No.2.

设置位置数据 PPL1，PPL2，与 PPL3 以决定在 No.1 机器人和 No.2 机器人中的公共框架坐标。在 No.1 机器人和 No.2 机器人中的公共框架坐标设置公共坐标。

2）Set the common frame coordinate in robot No.1 and robot No.2.

在 No.1 机器人和 No.2 机器人中设置的公共框架坐标。

Program　程序样例：

```
PFR1=Fram（PPL1, PPL2, PPL3）　'——Calculate the origin data of common frame
                                    of robot No.1./ 计算 No.1 机器人的公共框架
                                    的原点数据。

PFR2=Fram（PPL1, PPL2, PPL3）　'——Calculate the origin data of common frame
                                    of robot No.2./ 计算 No.2 机器人的公共框架
                                    的原点数据。
```

Chapter 8 Advanced Functions

3) Use the setting of the robot No.1 base coordinate for robot No.2 base coordinate (setting on robot No.2 only).

对于 No.2 机器人基本坐标，设置 No.1 机器的基本坐标（仅仅设置 No.2 机器人）。

Program 程序：

Share the setting of the robot No.1 base coordinate with robot No.2.

通过设置 No.2 机器人基本坐标共享设置 No.1 机器的基本坐标。

```
PBTMP=PFR2*Inv (PFR1)
PBASE=Inv (PBTMP)
Base PBASE                    '——Set the base coordinate./ 设置基本坐标。
```

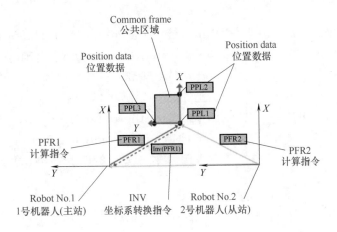

图 8-36 调整 1

（5）Adjustment 2: setting of the common tool 设置公共抓手（图 8-37）

Use the robot program BFRM.prg to set the common tool in robot No.1 and robot No.2.Establish the tool position at midpoint between the workpiece grasp positions for robot No.1 and robot No.2.

使用机器人程序 BFRM.prg 设置 No.1 机器人和 No.2 机器人的公共抓手。建立的抓手位置在 No.1 机器人和 No.2 机器人夹持工件的中点位置。

Outline of setting procedure 设置方法：

1) Teach the workpiece grasp positions for robot No.1 and robot No.2（PPK1, PPK2）.

示教获得 No.1 机器人和 No.2 机器人夹持工件位置（PPK1，PPK2）。

PPK1: workpiece picking position of robot 1.PPK2: workpiece picking position of robot 2.

PPK1：No.1 机器人夹持工件位置。PPK2：No.2 机器人夹持工件位置。

2) Set the tool coordinate of robot 1.

设置 No.1 机器人抓手坐标。

Program 程序：

PBT=(PPK1+PPK2)/2　'——Determine the midpoint by calculation between the workpiece removing positions for robot No.1 and robot No.2./ 根据 No.1 机器人和 No.2 机器人搬运工件的位置计算中点位置。

PTL=Inv(PPK1)*PBT　'——Determine the common tool by calculation in robot No.1 and robot No.2 Tool PTL./ 根据 No.1 机器人和 No.2 机器人的抓手位置计算公共抓手位置。

3）Set the tool coordinate of robot 2.
设置 No.2 机器人的抓手坐标。
Program 程序：

PBT=(PPK1+PPK2)/2　'——Determine the midpoint by calculation between the workpiece removing positions for robot No.1 and robot No.2./ 根据 No.1 机器人和 No.2 机器人搬运工件的位置计算中点位置。

PTL=Inv(PPK2)*PBT　'——Determine the common tool by calculation in robot No.1 and robot No.2 Tool PTL./ 根据 No.1 机器人和 No.2 机器人的抓手位置计算公共抓手位置。

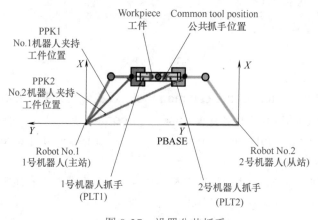

图 8-37　设置公共抓手

（6）Adjustment 3　调节 3
Teaching and parameter setting of the workpiece transport destination.
工件传送目标的示教和参数设置（图 8-38）。
Use the robot program 1.prg for robot No.1 to teach the workpiece transport destination (to robot No.1 only).
使用机器人程序"program 1.prg"对 1 号机器人示教工件传送目标位置。
Set the parameter to enable each extended function for robot No.1 and robot No.2.
通过设置参数扩展 1 号机器人和 2 号机器人的功能。

Outline of setting procedure　设置方式概述：

1）Teach the workpiece transport destination to robot No.1（P1, P2）.
示教 1 号机器人的工件传送目标位置（P1, P2）。

2）Set the parameter to enable each extended function for robot No.1 and robot No.2.
为 1 号机器人和 2 号机器人设置参数以扩展其功能。

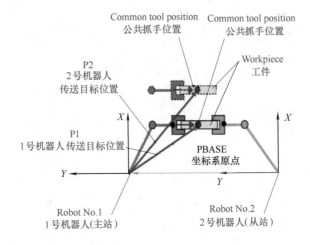

图 8-38　示教与设置

（7）Parallel straight motion by robot No.1　1 号机器人平行直线移动（图 8-39）

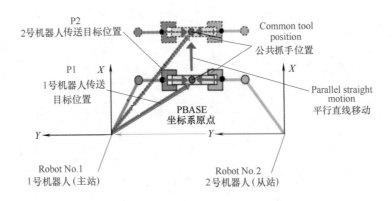

图 8-39　平行直线运动

（8）Rotating motion by robot No.1　1 号机器人的旋转运动（图 8-40）

Robot No.1 and robot No.2 rotate the workpiece about the common tool position as a center of rotation.

1 号机器人和 2 号机器人以公共抓手位置为旋转中心转动工件。

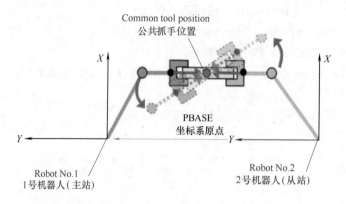

图 8-40　旋转运动

Chapter 9
Tracking Control
跟踪控制

9.1　Tracking systems　跟踪系统

9.1.1　Configuration example of conveyor tracking systems
　　　　传输线跟踪系统

（1）Conveyor tracking systems　传输线跟踪系统1（图 9-1）

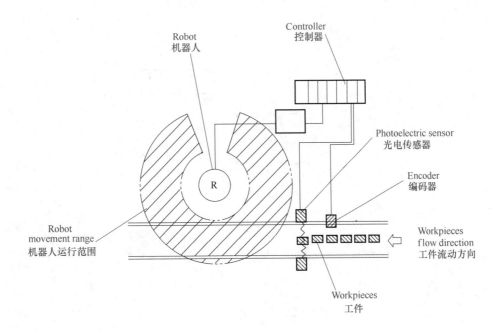

图 9-1　传输线跟踪系统 1

（2）Conveyor Tracking Systems2　传输线跟踪系统 2（图 9-2）

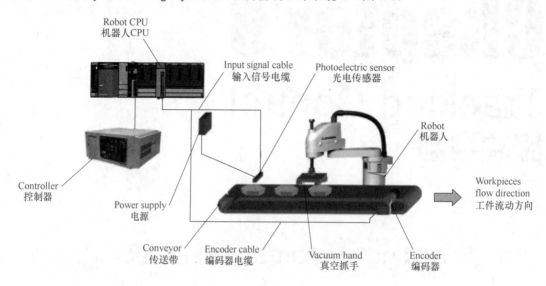

图 9-2　传输线跟踪系统 2

9.1.2　Configuration example of vision tracking systems 视觉跟踪系统

（1）Vision tracking systems 1　视觉跟踪系统 1（图 9-3）

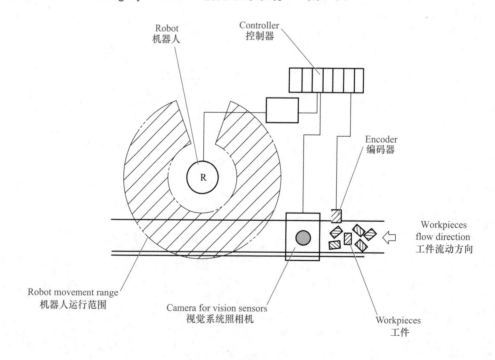

图 9-3　视觉跟踪系统 1

(2) Vision tracking systems 2　视觉跟踪系统2（图9-4）

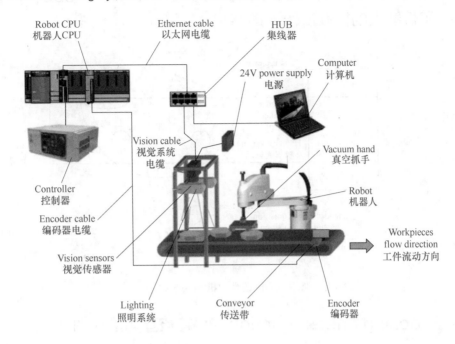

图 9-4　视觉跟踪系统2

9.1.3　Measures against the noise　抗电磁干扰（图9-5）

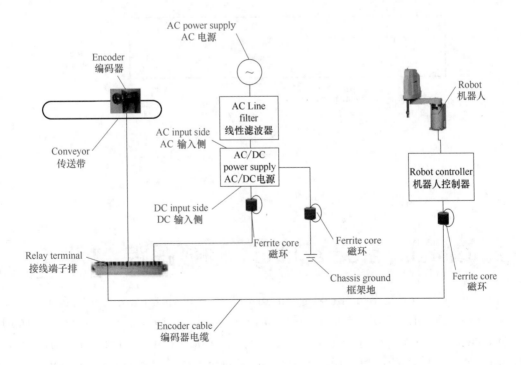

图 9-5　抗电磁干扰

9.1.4 Calibration operation for conveyor and robot 传送带和机器人的校准操作（图 9-6）

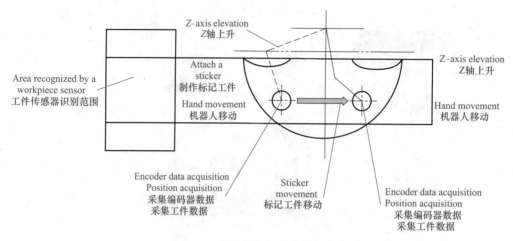

图 9-6 传送带和机器人的操作

9.1.5 Tracking check function 跟踪核查功能（图 9-7）

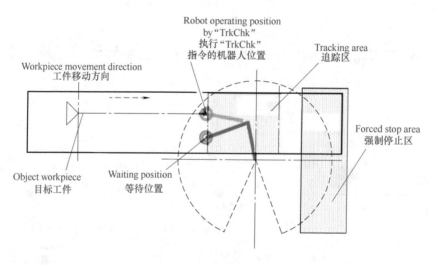

图 9-7 跟踪核查功能

9.2 Circular arc tracking 圆弧跟踪

What is the circular arc tracking function? The circular arc tracking function allows robots to follow workpiece on a turntable and a circular arc conveyor. With this function, it becomes possible to transport line up and process workpieces without having to stop the conveyor. It also eliminates the need for mechanical fixtures and so forth required to fix workpiece positions.

什么是圆弧追踪功能？"圆弧追踪功能"即机器人追踪在旋转工作台和圆弧型传送带上的工件的功能。这种方式不需要停止传送带也可以处理传送带上的工件，同时也不需要更多的机械夹具。

（1）Straight conveyor　直线型传送带（图9-8）

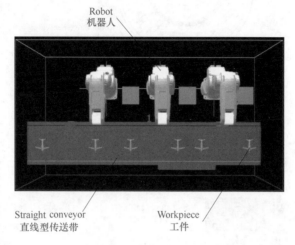

图9-8　直线型传送带

（2）Circular arc conveyor　圆弧型传送带（图9-9）

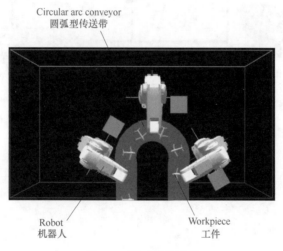

图9-9　圆弧型传送带

（3）Tracking area of straight conveyor　直线传送带的追踪区（图9-10）

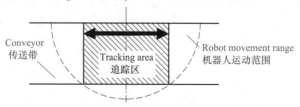

图9-10　直线传送带的追踪区

(4) Tracking area of circular arc conveyor　圆弧追踪区（图9-11）

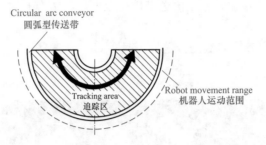

图9-11　圆弧跟踪区

(5) Configuration example of Q type　Q型系统构成（图9-12）

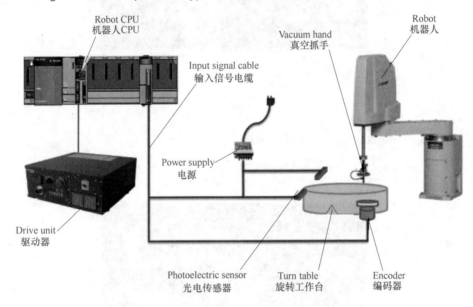

图9-12　Q型系统构成

Chapter 10
Additional Axis
附加轴

10.1 Outline 概述

The additional axis function is a function which uses the general-purpose servo amplifier of the corresponding servomotors in order to allow the plural above servomotors to be controlled from the robot controller.

附加轴功能是由机器人控制通用伺服放大器以驱动伺服电机的功能。

（1）Travel axis system 行走轴系统（图 10-1）

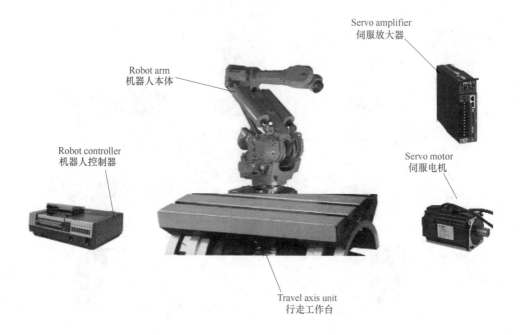

图 10-1 行走轴系统

(2) Rotation table system　旋转台系统（图 10-2）

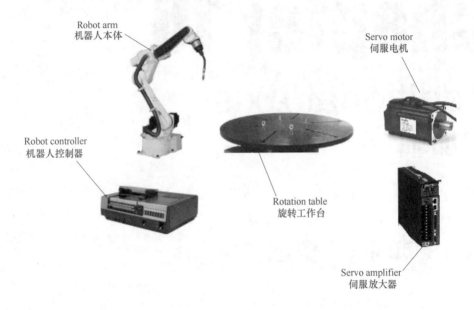

图 10-2　旋转台系统

(3) Multi-axis system　多轴系统（图 10-3）

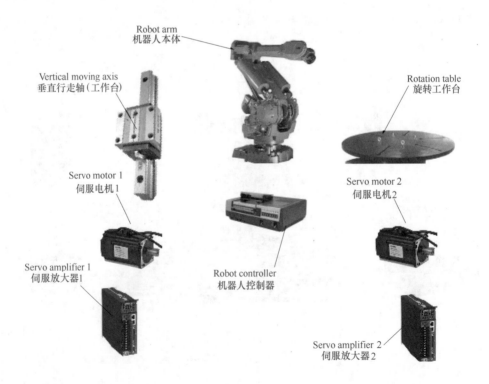

图 10-3　多轴系统

Chapter 10　Additional Axis

(4) Connection 1　连接1（图10-4）

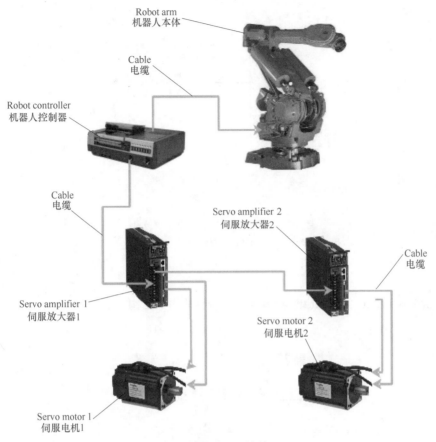

图10-4　连接

(5) Connection 2　连接2（图10-5）

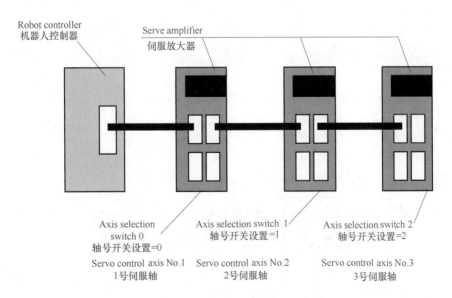

图10-5　控制序号和伺服参数

（6）Internal structure of travel station　行走台的内部结构（图10-6）

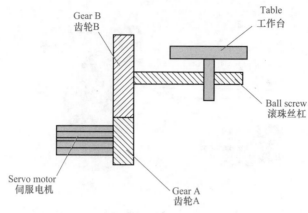

图 10-6　行走台的内部结构

（7）Internal structure of rotating table　旋转台的内部结构（图10-7）

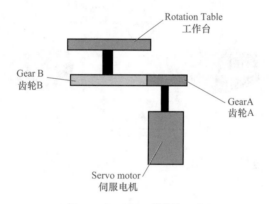

图 10-7　旋转台的内部结构

（8）Example of arc interpolation　圆弧插补样例（图10-8）

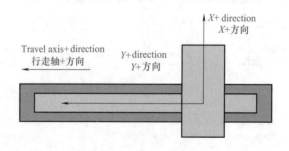

图 10-8　圆弧插补样例

Chapter 10　Additional Axis

(9) Handling workpiece　搬运工件（图 10-9）

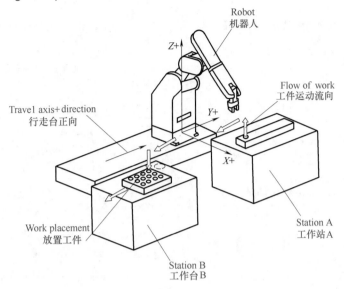

图 10-9　搬运工件

(10) Position variable　位置变量（图 10-10）

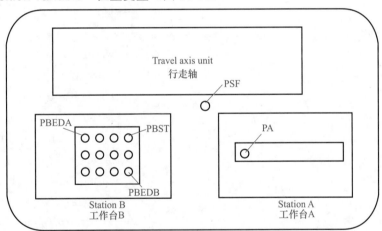

图 10-10　位置变量

表 10-1　位置变量的说明

Mechanism name 名称	Position variable name 变量名称	Explanation 说明
Robot arm 机器人本体	PSF	Safe position　安全位置
	PA	Position where works are unloaded from station A 工作台 A 的工件卸放位置
	PBST	Position where works are loaded to station B（start position of pallet） 工件被夹持到工作台 B 位置（货盘起点位置）
	PBEDA	Position where works are loaded to station B（end position of pallet A） 工件被夹持到工作台 B 位置（货盘 A 排终点位置）
	PBEDB	Position where works are loaded to station B（end position of pallet B） 工件被夹持到工作台 B 位置（货盘 B 排终点位置）

Program of mechanism number 1（program name：1）1号机器人程序：

```
1  Def Plt 1, PBST, PBEDA, PBEDB, 4, 3, 2
```
'——Definition of pallet number 1./ 定义 1 号托盘。

```
2  Mov PSF
```
'——Move to safe position./ 移动到 PSF "安全位置点"。

```
3  HOpen 1
```
'——Open the hand1./ 张开抓手 1。

```
4  M1=1
```
'——M1 is used for counter./ 定义 M1 为计数器。

```
5  *W1
```
'——标签。

```
6  If M_In（11）=0 Then GoTo *W1
```
'——Waits for the transport of a work./ 如果输入信号（11）=0，就返回 *W1 行，即等待搬运工件。

```
7  M_Out（11）=0
```
'——Transporting a work./ 如果输出信号（11）=0，就表示在搬运工件。

```
8  *LOOP
```
'——循环标志。

```
9  Mov PA, -50
```
'——Moves to the position of 50mm back from work unloaded position./ 运动到 PA 上部 50mm。

```
10  Mvs PA
```
'——Moves to the position where work is unloaded./ 运动到卸放工件位置 PA。

```
11  HClose 1
```
'——Close the hand1./ 1 号抓手夹持工件。

```
12  Dly 0.5
```
'——Waits for 0.5 s./ 等待 0.5s。

```
13  Mvs PA, -50
```
'——Moves to the position of 50 mm back from work unloaded position./ 运动到 PA 上部 50mm。

```
14  PB=（Plt 1, M1）
```
'——Calculates the position in the pallet number 1 indicated by M1./ 根据 M1 变量计算的工件放置点位置。

```
15  Mov PB, -50
```
'——Moves to the position 50 mm back from the work placing position./ 运动到 PB 上部 50mm。

16 Mvs PB	'——Moves to the work placing position./ 运动到卸放工件位置PB。
17 HOpen 1	'——Open the hand1./ 张开抓手1。
18 Dly 0.5	'——Waits for 0.5 s./ 等待0.5s。
19 Mvs PB, -50	'——Moves to the position 50 mm back from the work placing position./ 运动到PB 上部50mm。
20 M1=M1+1	'——Advances the counter./M1 做变量运算。
21 If M1<=12 Then *LOOP	'——Loops as many as the number of works./ 如果M1≤12，则继续循环运行。
22 M_Out（11）=1	'——Work full./ 发出输出（11）=1 信号，工件装满托盘。
23 End	'——程序结束。

（11）Rotation Table work case　旋转工作台工作案例（图 10-11）

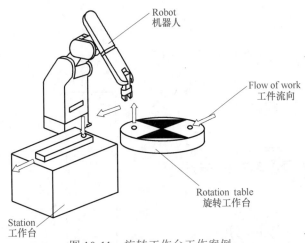

图 10-11　旋转工作台工作案例

（12）Position variables used in programming　编程使用的位置变量（表 10-2、图 10-12）

表 10-2　编程使用的位置变量

Mechanism name 名称	Position variable name 变量名称	Explanation 说明
Robot arm 机器人本体	P1SF	Safe position　安全位置
	P11	Position where works are unloaded from rotation axis（mechanism No.2） 从旋转工作台过来的工件卸放位置
	P12	Position where works are loaded to station 在工作站上的工件夹持位置
	P231	Position to which the work is to be transported 工件传送位置
	P232	Position from which the standard robot unloads the work 卸放工件位置

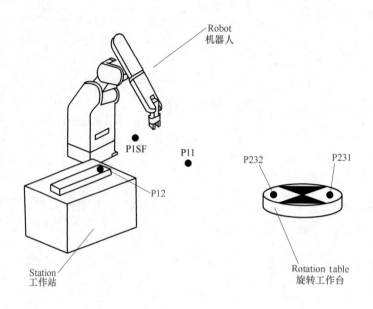

图 10-12 编程使用的位置变量

（13）Flow of work　工作流程（图 10-13）

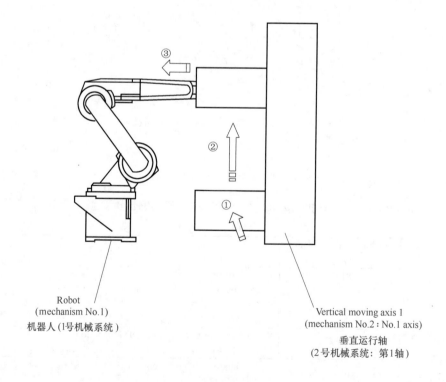

图 10-13　工作流程

10.2 Additional axis function 附加轴功能

Connect servo driver 连接伺服驱动器（图 10-14）：

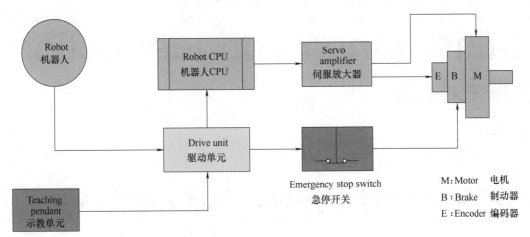

图 10-14　连接伺服驱动器

Chapter 11
Maintenance and Inspection
维护和保养

11.1 Robot arm structure 机器人构造

11.1.1 Structure for horizontal multi-joint type robot 水平型机器人的构造

(1) Structure for horizontal multi-joint type robot (3kg) 水平型机器人3kg系列（图11-1）

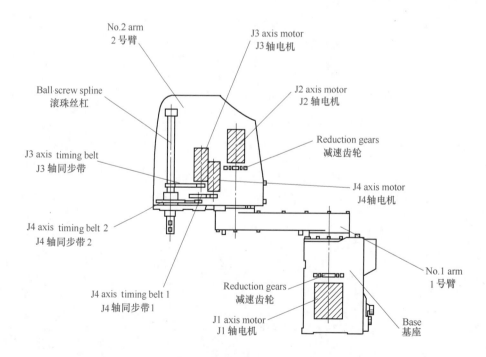

图11-1 机器人本体结构

Chapter 11 Maintenance and Inspection

（2）Structure for horizontal multi-joint type robot（6～20kg） 水平型机器人6～20kg系列（图11-2）

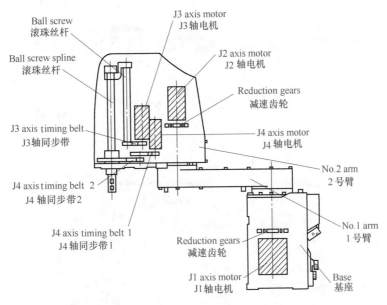

图 11-2 机器人本体结构 2

11.1.2 Structure for 6-axis robot 6 轴机器人的构造

（1）Structure for 6-axis robot 1 6 轴机器人构造 1（图 11-3）

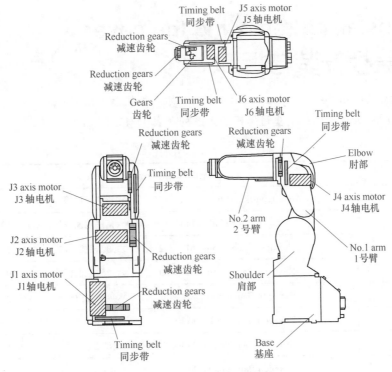

图 11-3 垂直型机器人构造 1

（2）Structure for 6-axis robot 2　6 轴机器人构造 2（图 11-4）

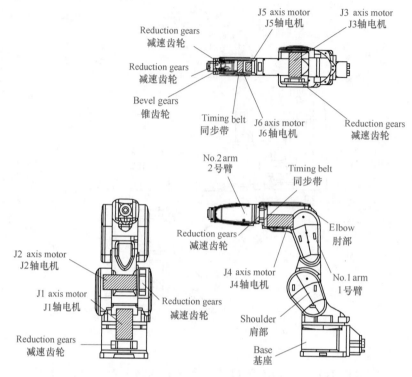

图 11-4　垂直型机器人构造 2

11.2　Installing/removing the cover　盖板安装与拆卸

（1）Horizontal multi-joint type robot series　水平型机器人系列（图 11-5）
（2）Vertical robot series　垂直型机器人系列（表 11-1、图 11-6）

表 11-1　有关符号的说明

Symbols 符号	Installation screws	安装螺栓
（a）	Hexagon flange bolt, M4×12	六角法兰面螺栓, M4×12
（b）	Hexagon flange bolt, M4×12	六角法兰面螺栓, M4×12
（c）	Hexagon flange bolt, M4×12	六角法兰面螺栓, M4×12
（d）	Hexagon flange bolt, M4×20	六角法兰面螺栓, M4×20
（e）	Hexagon flange bolt, M4×12	六角法兰面螺栓, M4×12
（f）	Hexagon flange bolt, M4×8	六角法兰面螺栓, M4×8
（g）	Hexagon flange bolt, M4×8	六角法兰面螺栓, M4×8
（h）	Hexagon flange bolt, M4×12	六角法兰面螺栓, M4×12

Chapter 11　Maintenance and Inspection

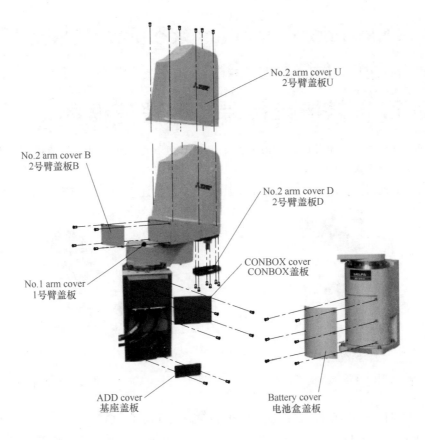

图 11-5　水平型机器人的盖板拆卸

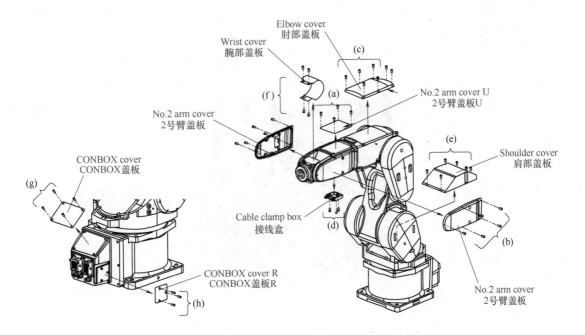

图 11-6　垂直型机器人的盖板拆卸

11.3 Inspection, maintenance and replacement of timing belt 同步带检查保养及张紧度调整

（1）Measurement by the sound wave type belt tension gauge　使用声波型同步带张力检测仪进行张力测量（图11-7）

（2）Measurement by the push-pull gauge　使用按压型检测仪测量压力（图11-8）

图11-7　使用声波型同步带张力检测仪进行张力的检测　　图11-8　使用按压型检测仪测量压力

11.3.1 Inspection and maintenance of timing belt for horizontal multi-joint type robot 水平型机器人同步带的检测和维护

（1）Adjustment and test of tension force of timing belt for J3 axis　J3轴同步带张紧力的调整和检测（图11-9）

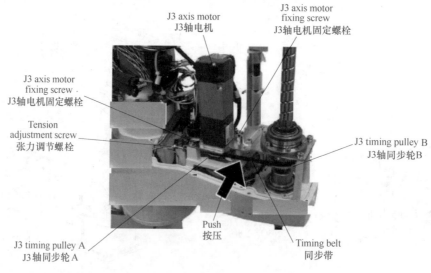

图11-9　J3轴同步带张紧力的调整和检测

(2) Inspection and maintenance of J4 axis timing belt　J4 轴同步带的检查与调整（图 11-10）

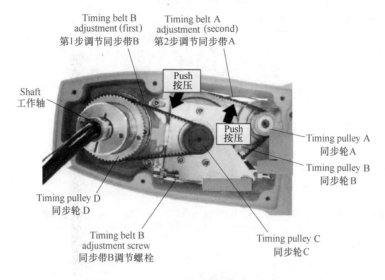

图 11-10　J4 轴同步带的检查与调整

11.3.2　Inspection and maintenance of timing belt for 6-axis robot　6 轴机器人同步带保养维护

(1) Inspection and maintenance of timing belt for J1 axis　J1 轴同步带保养维护（图 11-11）

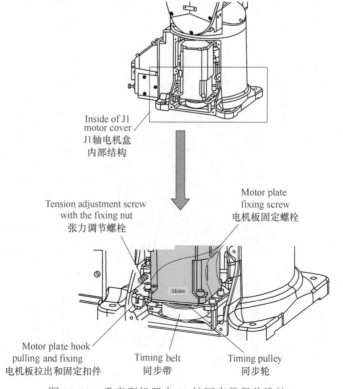

图 11-11　垂直型机器人 J1 轴同步带保养维护

（2）Inspection and maintenance of timing belt for J3 axis　J3 轴同步带保养维护（图 11-12）

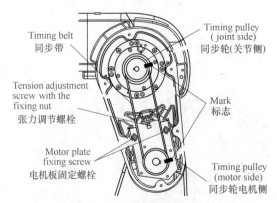

图 11-12　垂直型机器人 J3 轴同步带保养维护

（3）Inspection and maintenance of timing belt for J4 axis 1　J4 轴同步带保养维护 1（图 11-13）

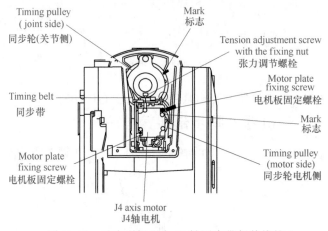

图 11-13　垂直型机器人 J4 轴同步带保养维护 1

（4）Inspection and maintenance of timing belt for J4 axis 2　J4 轴同步带保养维护 2（图 11-14）

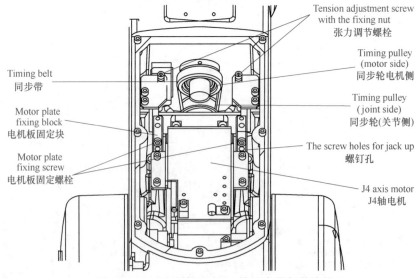

图 11-14　垂直型机器人 J4 轴同步带保养维护 2

（5）Inspection and maintenance of timing belt for J5 axis　J5轴同步带保养维护（图11-15）

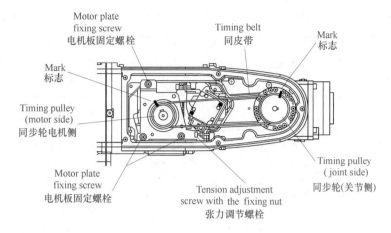

图 11-15　垂直型机器人 J5 轴同步带保养维护

（6）Inspection and maintenance of timing belt for J6 axis　J6轴同步带保养维护（图11-16）

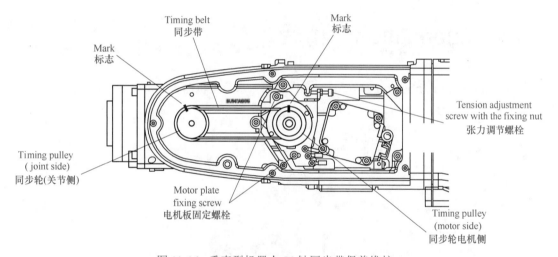

图 11-16　垂直型机器人 J6 轴同步带保养维护

11.3.3　Tension adjustment of timing belt　同步带的张紧调整（图 11-17）

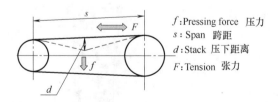

图 11-17　皮带张紧力的调整和检测

Table 11-2 is the preset value and adjustment value in the sound wave type belt tension gauge.

表 11-2 是同步带的压力和张力调节标准。

表 11-2 同步带的压力和张力调节标准

Axis 轴号	Press force parameters 压力参数			Standard tension 标准张力 /N	
	M/g·m^{-1}	W/mm·r^{-1}	S/mm	New blet 新同步带	Used blet 使用过的同步带
J1	4.0	15	107.5	129.6～158.4	86.4～105.6
J3	2.5	9	178.5	59.4～72.6	39.6～48.4
J4	2.5	6	54.9	39.2～47.9	26.1～31.9
J5	2.5	6	150.0	39.2～47.9	26.1～31.9
J6	2.5	4	96.0	27.0～33.0	18.0～22.0

11.4 Lubrication 润滑

(1) Lubrication position of horizontal robot 水平型机器人润滑位置（图 11-18）

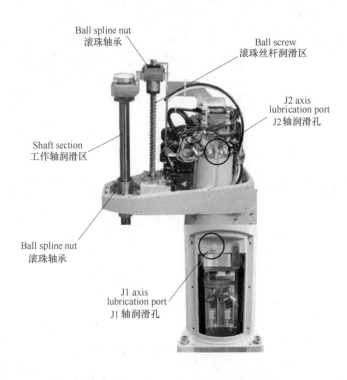

图 11-18 水平型机器人润滑位置

（2）Lubrication position of vertical robot1　垂直型机器人润滑位置1（图11-19）

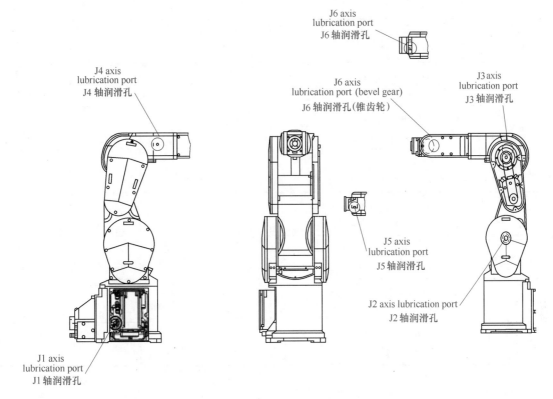

图 11-19　垂直型机器人润滑位置 1

（3）Lubrication position of vertical robot 2　垂直型机器人润滑位置 2（图 11-20）

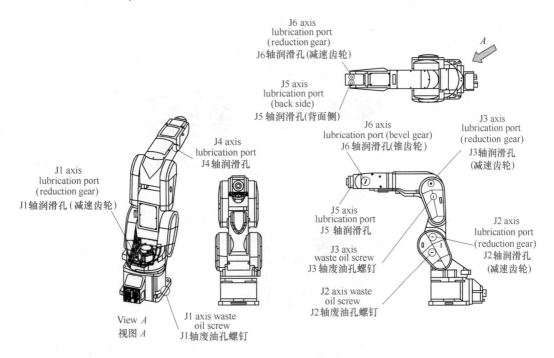

图 11-20　垂直型机器人润滑位置 2

（4）Lubrication specifications　润滑规范（表 11-3）

表 11-3　润滑规范

No. 序号	Parts to be lubricated 润滑部件	Lubrication interval 润滑时间间隔 /h	Lubrication amount 润滑油量 /g
1	J1 axis reduction gears J1 轴减速齿轮	20000	255
2	J2 axis reduction gears J2 轴减速齿轮		251
3	J3 axis reduction gears J3 轴减速齿轮		150
4	J4 axis reduction gears J4 轴减速齿轮	24000	7
5	J5 axis reduction gears J5 轴减速齿轮		3
6	J6 axis reduction gears J6 轴减速齿轮		2
7	J6 axis bevel gears J6 轴锥齿轮		1.5

11.5　Replacing the battery　更换电池

（1）Batteries in the controller　控制器内的电池（图 11-21）

图 11-21　控制器内的电池

Chapter 11　Maintenance and Inspection

（2）Replacement of batteries for horizontal robots　水平型机器人更换电池（图 11-22）

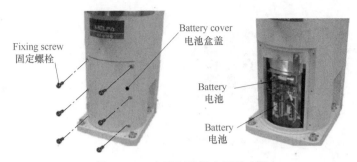

图 11-22　水平型机器人更换电池

（3）Replacement of batteries for vertical robots　垂直型机器人更换电池（图 11-23）

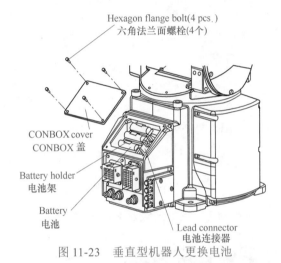

图 11-23　垂直型机器人更换电池

11.6　The check of the filter, cleaning, exchange 过滤窗的检查、清洗及更换

（1）Structure of filter window　过滤窗的结构（图 11-24）

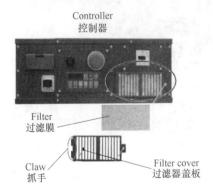

图 11-24　过滤窗的检查、清洗及更换

（2）Replacement of bellows1　更换波纹管1（图11-25）

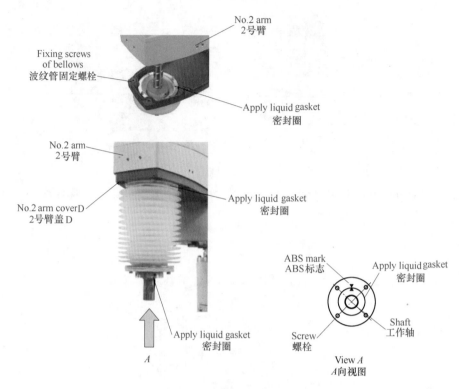

图11-25　更换波纹管1

（3）Replacement of bellows 2　更换波纹管2（图11-26）

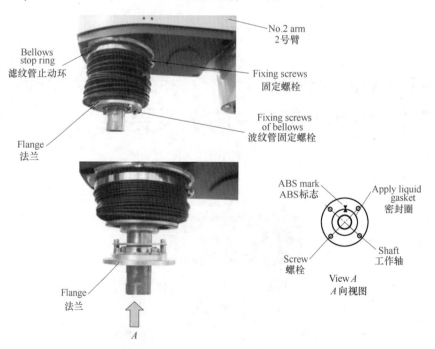

图11-26　更换波纹管2

11.7 Overhaul 大修（图 11-27）

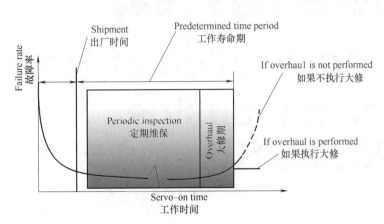

图 11-27　大修

Chapter 12

The Application of Robots in Welding Industries
机器人在焊接行业中的应用

Welding robots are industrial robots engaged in welding (including cutting and spraying). According to the International Organization for Standardization (ISO) definition of industrial robot belongs to the standard of welding robot, industrial robot is a kind of multipurpose, reprogrammable, automatically controled manipulator, with three or more programmable Exes, used in industrial automation field.In order to adapt to different uses, the mechanical interface of the last shaft of the robot, usually a connecting flange, can be connected to different tools or called end-effector. Welding robots are equipped with welding electrode holders or welding (cutting) guns on the final shaft flange of industrial robots, enabling them to weld, cut or spray. Fig.12-1 shows a welding robot.

焊接机器人是从事焊接（包括切割与喷涂）的工业机器人。根据国际标准化组织（ISO）关于标准焊接机器人的定义，工业机器人是一种多用途的、可重复编程的自动控制操作机（manipulator），具有三个或更多可用程序控制的轴，用于工业自动化领域。为了适应不同的用途，机器人最后一个轴的机械接口，通常是一个连接法兰，可接装不同工具或称"末端执行器"。焊接机器人就是在工业机器人的工作轴法兰上装接焊钳或焊（割）枪，使之能进行焊接、切割或热喷涂。图 12-1 所示是一种焊接机器人。

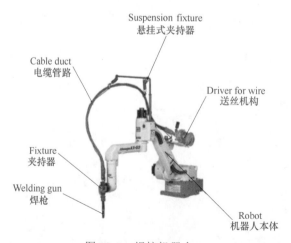

图 12-1　焊接机器人

Chapter 12 The Application of Robots in Welding Industries

12.1 Composition and structure
组成结构

Welding robot mainly consists of two parts: robot and welding equipment. The robot is composed of robot body and control cabinet (hardware and software). The welding equipment, take arc welding and spot welding as an example, is made up of welding power supply (including its control system), wire feeder (arc welding), welding gun (electrode holder) and other parts. There should also be sensing system for intelligent robot, such as laser or camera sensor and control device. Fig.12-2 show the basic components of arc welding robot and spot welding robot.

焊接机器人主要包括机器人和焊接设备两部分。机器人由机器人本体和控制柜（硬件及软件）组成。而焊接装备，以弧焊及点焊为例，则由焊接电源（包括其控制系统）、送丝机（弧焊）、焊枪（钳）等部分组成。对于智能机器人还有传感系统，如激光或摄像传感器及其控制装置等。图 12-2 所示是弧焊机器人和点焊机器人的基本组成。

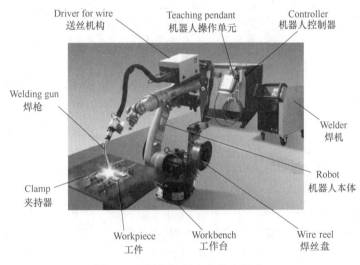

图 12-2　焊接机器人的基本组成

Welding robots are basically joint robots, most of which have six axes. The 1, 2, and 3 axes send the tool to different spatial locations, while the 4, 5, and 6 axes are used to adjust the shape of the tool.

焊接用机器人基本上都属于关节型机器人，大部分有 6 个轴。其 1、2、3 轴将工具送到不同的空间位置，而 4、5、6 轴用于调整工具的形位。

The robot is driven by a servo motor through cycloidal needle wheel (RV) reducer (1~3 axis) and harmonic reducer (1~6 axis). AC servo motor is used. Because AC motor does not have carbon brush, the motion characteristic is good, add (subtract) the speed is quick. The tool center point (TCP) of a lightweight robot with a load of less than 16kg can reach a maximum motion speed of more than

3m/s with accurate positioning and small vibration. At the same time, the controller of the robot uses a 32-bit microcomputer and a new algorithm, so that the robot has the function of self-optimizing the path, and the running trajectory is more in line with the teaching trajectory.

机器人各轴都做回转运动，采用伺服电机通过摆线针轮（RV）减速器（1～3轴）及谐波减速器（1～6轴）驱动。采用交流伺服电机。由于交流电机没有碳刷，运动特性好，加（减）速度快。负载16kg以下的轻型机器人的工具中心点（TCP）最高运动速度可达3m/s以上，定位准确，振动小。同时，机器人的控制器使用32位微机和新的算法，使机器人具有自行优化路径的功能，运行轨迹更符合示教轨迹。

12.2　Feature　特点

The requirement of spot welding robot is not very high. Because spot welding only requires point control, there is no strict requirement for the moving track of welding electrode holders between points, which is why the robot can only be used for spot welding in the earliest time. The robot used for spot welding should not only have enough load capacity, but also be quick, smooth and accurate when shifting from point to point, so as to reduce the time of shifting and improve work efficiency.

点焊对焊接机器人的要求不是很高。因为点焊只需点位控制，对焊钳在点与点之间的移动轨迹没有严格要求，这也是机器人最早用于点焊的原因。点焊用机器人不仅要有足够的负载能力，而且在点与点之间移位时速度要快捷，动作要平稳，定位要准确，以减少移位的时间，提高工作效率。

How much load capacity the spot-welding robot needs depends on the type of electrode holder used. For the use of transformer-separated welding electrode holders, using 30~45kg load of the robot. However, due to the long secondary cable, the power consumption of the electrode holder is large, which is not conducive to the welding of the electrode holder into the workpiece; moreover, the cable oscillates with the robot's movement, and the cable is damaged faster. Therefore, the current use of integrated welding electrode holders.The mass of the electrode holder together with the transformer is about 70kg.Because the robot should have enough load capacity, can quickly send the welding electrode holders to the space position for welding, generally choose 100~150kg load heavy robot.

点焊机器人需要有多大的负载能力，取决于所用的焊钳形式。对于与变压器分离的焊钳，使用30～45kg负载的机器人。但是，这种焊钳由于二次电缆线长，电能损耗大，不利于将焊钳伸入工件内部焊接；且电缆随机器人运动而不停摆动，电缆的损坏较快。因此，目前多采用一体式焊钳。这种焊钳连同变压器质量在70kg左右。由于机器人要有足够的负载能力，能快速将焊钳送到空间位置进行焊接，一般都选用100～150kg负载的重型机器人。

Chapter 12　The Application of Robots in Welding Industries

12.3　Structure design　结构设计

The welding robot works in the environment of narrow space. In order to ensure that the robot can track the automatic welding of welds according to the information of arc sensor, it is required that the designed robot should have compact structure, flexible movement and stable operation. The robot mechanism is divided into three parts: wheeled mobile platform, welding torch regulating mechanism and arc sensor. Among them, due to its large inertia and slow response, the wheeled mobile platform mainly tracks the welding seam roughly, the welding torch adjustment mechanism is responsible for accurate tracking of the welding seam, and the arc sensor completes the real-time identification of the welding seam deviation. In addition, the robot controller and motor driver are integrated on the robot mobile platform to make it smaller. At the same time, in order to reduce the influence of dust on the moving parts in the harsh welding environment, the fully enclosed structure is adopted to improve the system reliability. Fig.12-3 shows the basic components of arc welding robot.

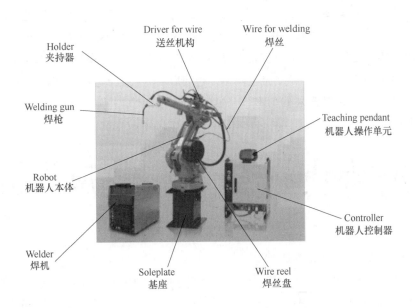

图 12-3　弧焊机器人的基本组成部分

焊接机器人是在空间狭窄的环境下工作，为了保证机器人能根据电弧传感器的信息跟踪焊缝自动焊接，要求所设计的机器人应该结构紧凑、移动灵活且工作稳定。机器人机构分为轮式移动平台、焊炬调节机构和电弧传感器三部分。其中，轮式移动平台由于其惯性大，响应慢，主要对焊缝进行粗跟踪，焊炬调节机构负责焊缝精确跟踪，电弧传感器完成焊缝偏差实时识别。另外，机器人控制器和电机驱动器集成安装于机器人移动平台上，使其体积更小。同时，为了减少恶劣焊接环境下粉尘对运动部件影响，采用全封闭式结构，提高其系统可靠性。图 12-3 所示是弧焊机器人的基本组成部分。

12.4　Equipment　装备

　　Spot welding robot welding equipment, because of the use of integrated welding electrode holders, welding transformers installed in the back of the welding electrode holders, so the transformer must be as small as possible. For transformers with small capacity, 50Hz power frequency AC can be used, while for transformers with large capacity, inverter technology has been adopted to change 50Hz power frequency AC into 600~700Hz AC, so as to reduce the volume of the transformer.After the change of voltage, it can be directly welded with 600 ~ 700Hz AC, and can also be further rectified with DC welding. Welding parameters are adjusted by the timer, spot welding robot's welding electrode holders, usually with pneumatic welding electrode holders, pneumatic welding electrode holders between the two electrodes generally only two strokes. Moreover, once the electrode pressure is adjusted, it cannot be changed at will.

　　点焊机器人的焊接装备，由于采用了一体化焊钳，焊接变压器装在焊钳后面，所以变压器必须尽量小型化。对于容量较小的变压器可以用 50Hz 工频交流，而对于容量较大的变压器，已经开始采用逆变技术把 50Hz 工频交流变为 600～700Hz 交流，使变压器的体积减小。变压后可以直接用 600～700Hz 交流电焊接，也可以再进行二次整流，用直流电焊接。焊接参数由定时器调节，点焊机器人的焊钳，通常用气动的焊钳，气动焊钳两个电极之间的开口度一般只有两级冲程，而且电极压力一旦调定后是不能随意改变的。

12.5　Welding application　焊接应用

12.5.1　Workstation　工作站

　　This system is simplest if the workpiece is not in a position to be changed throughout the welding process. But in the actual production, more workpiece in the welding need to change the position, so that the welding seam is in the appropriate position.In this case, the transformer and the robot can move separately, that is, the robot rewelds after the transformer moves the workpiece;it can also be simultaneous motion, that is, the robot welding while the displacement machine moves the workpiece. At this time, the joint movement of the displacement machine and the robot can make the welding torch relative to the workpiece to meet both the weld trajectory and the welding speed and the position of the welding gun. At this time, the moving axis of the displacement machine has become a part of the robot, and the welding robot system can have as many as 7~20 axes.Fig.12-4 shows

Chapter 12 The Application of Robots in Welding Industries

a welding robot with a turnover device.

如果工件在整个焊接过程中无需变动位置，可以用夹具把工件固定位在工作台上，这种系统是最简单的。但在实际生产中，更多的工件在焊接时需要变动位置，使焊缝处在合适的位置下焊接。对于这种情况，变位机与机器人可以是分别运动，即变位机移动工件后机器人再焊接，也可以是同时运动，即变位机移动工件的同时，机器人进行焊接。这时变位机与机器人联合运动，使焊枪相对于工件的运动既能满足焊缝轨迹，又能满足焊接速度及焊枪位置的要求。这时变位机的运动轴已成为机器人的组成部分，这种焊接机器人系统可以多达7~20个轴。图12-4所示是带翻转工作台的焊接机器人。

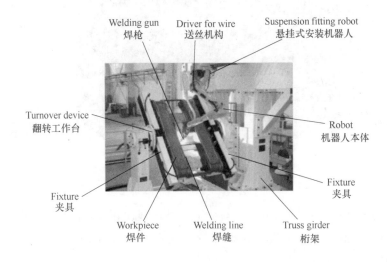

图 12-4 带翻转工作台的焊接机器人

12.5.2 Box welding robot workstation 箱体焊接机器人工作站

Box welding robot workstation is specially used in cabinet industry. Cabinet industry has large production, welding quality and size requirements.

箱体焊接机器人工作站专门用于箱柜行业。箱柜行业生产量大，焊接质量及尺寸要求高。

The box welding robot workstation is composed of arc welding robot, welding power supply, welding gun wire feeding mechanism, rotary double-position shifting machine, fixture and control system. The workstation is suitable for welding of all kinds of box workpieces and can realize automatic welding of various boxes by using different clamps in the same workstation. Fig.12-5 shows the robot workstation for box welding.

箱体焊接机器人工作站由弧焊机器人、焊接电源、焊枪送丝机构、回转双工位变位机、工装夹具和控制系统组成。该工作站适用于各式箱体类工件的焊接，在同一工作站内通过使用不同的夹具可实现多品种的箱体自动焊接。图12-5所示是箱体焊接机器人工作站。

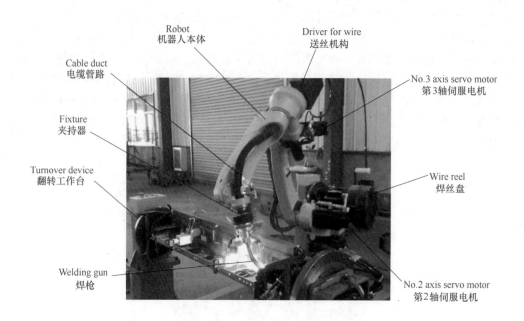

图 12-5　箱体焊接机器人工作站

As a result of the use of double-station shifting machine, at the same time of welding, other stations can be disassembled workpiece, greatly improve the welding efficiency. Because the MIG pulse transition or CMT cold metal transition welding process is adopted for welding, the heat input in the welding process is greatly reduced to ensure that the product will not be deformed after welding.

由于采用双工位变位机，焊接的同时，其他工位可拆装工件，极大地提高了焊接效率。由于采用了MIG脉冲过渡或CMT冷金属过渡焊接工艺方式进行焊接，使焊接过程中热输入量大大减少，保证产品焊接后不变形。

By adjusting the welding specification and robot welding posture, it can ensure the welding quality of the product is good, the welding seam is beautiful, especially for the stainless steel gas chamber with high requirements for sealing, the gas chamber after welding is guaranteed not to leak. By setting the variety selection parameters in the control system to replace the fixture, the automatic welding of multiple varieties of boxes can be realized. Fig.12-6 shows welding robot welding parts.

通过调整焊接规范和机器人焊接姿态，能够保证产品焊缝质量好，焊缝美观，特别对于密封性要求高的不锈钢气室，焊接后保证气室气体不泄漏。通过设置控制系统中的品种选择参数从而更换夹具，可实现对多个品种箱体的自动焊接。图12-6所示是焊接机器人焊接工件。

Chapter 12 The Application of Robots in Welding Industries

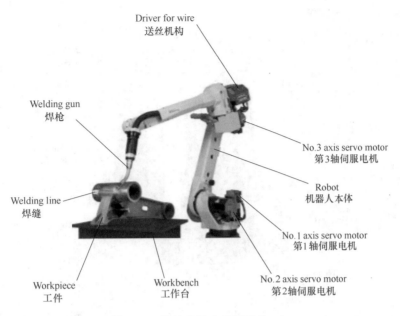

图 12-6　焊接机器人焊接工件

With different arc welding robots and transposition machines, the workstations can meet the welding requirements of all kinds of boxes with welding seam length around 2000mm. The welding speed is 3 ~ 10mm/s. According to the basic material of the box, different types of gas shielded welding are used in the welding process. Welding robot workstation can be used in power, electrical, mechanical, automotive and other industries.

用不同的弧焊机器人和变位机，工作站可以满足焊缝长度在 2000mm 左右的各类箱体的焊接要求。焊接速度 3~10mm/s，根据箱体基本材料，焊接工艺采用不同类型的气体保护焊。焊接机器人工作站可用于电力、电气、机械、汽车等行业。

12.5.3　Robot flexible laser welding machine　机器人柔性激光焊接机

Robot flexible laser welding machine for stainless steel welding, stainless steel chamber is a large deformation, sealing requirements of the box type workpiece. Robot flexible laser welding machine, laser generator units, water cooling unit by robots, laser scanning tracking system, the flexible displacement machine, tooling and fixture, security fence, dust collection device and other components of the control system, by setting the selection of parameters of the control system and replace the jig, can achieve a number of varieties of the automatic welding of the stainless steel chamber type of artifacts. Fig.12-7 shows the robot welding auto parts.

机器人柔性激光焊接机用于不锈钢焊接，不锈钢气室是变形量大、密封性要求高的箱体类工件。机器人柔性激光焊接机由机器人、激光发生器机组、水冷却机组、激光扫描跟踪系统、柔性变位机、工装夹具、安全护栏、吸尘装置和控制系统等组成，通过设置控制系统中的品种选择参数并更换工装夹具，可实现多个品种的不锈钢气室类工件的自动焊接。图 12-7

所示是机器人焊接汽车零部件。

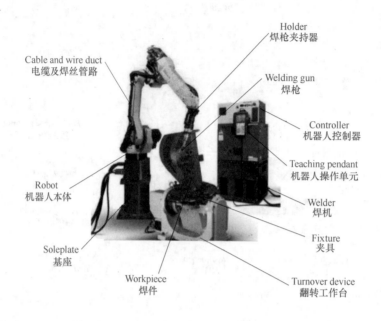

图 12-7　机器人焊接汽车零部件

12.5.4　Shaft welding robot workstation　轴类焊接机器人工作站

　　The shaft welding robot workstation is composed of arc welding robot, welding power supply, welding gun wire feeding mechanism, rotary double-position shifting machine, fixture and control system. Workstations are used for the welding of axle-type workpieces. In the same workstations, various types of shaft-type parts can be welded automatically by using different clamps. As a result of the use of double-station shifting machine, while welding, other stations can be disassembled workpiece, greatly improved the efficiency.

　　轴类焊接机器人工作站由弧焊机器人、焊接电源、焊枪送丝机构、回转双工位变位机、工装夹具和控制系统组成。工作站用于轴类工件的焊接，在同一工作站内通过使用不同的夹具可实现多品种的轴类零件自动焊接。由于采用双工位变位机，焊接的同时，其他工位可拆装工件，极大地提高了效率。

12.5.5　Robot welding stud workstation　机器人焊接螺柱工作站

　　Robot welding stud workstations can weld studs of different specifications to complex workpieces. Workstation by robot, welding power supply, automatic screw machine, welding gun, displacement machine, jig, automatic gun change device, automatic detection software, control system and safety guardrail, etc, using automatic screw machine send stud to robot automatic welding gun, by programming the robot path along the rules work, different specifications of the stud welding on the workpiece. The stud can be welded to the workpiece by energy storage welding or pull arc welding to ensure the welding precision and strength. Fig.12-8 shows the welding robot with wire feeding device.

机器人焊接螺柱工作站可将不同规格螺柱焊接到复杂的工件上。工作站由机器人、焊接电源、自动送钉机、焊枪、变位机、工装夹具、自动换枪装置、自动检测软件、控制系统和安全护栏等组成，通过自动送钉机将螺柱送到机器人自动焊枪内，通过编程使机器人沿规的路径工作，将不同规格的螺柱焊接到工件上。可以采用储能焊接或拉弧焊接将螺柱焊接到工件上，保证焊接精度和焊接强度。图 12-8 所示是带送丝装置的焊接机器人。

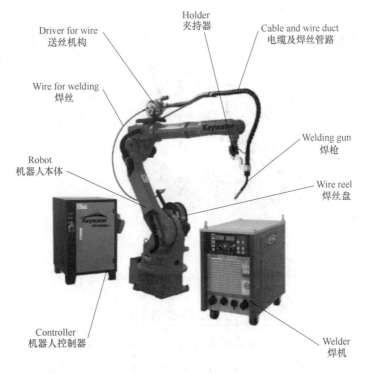

图 12-8　带送丝装置的焊接机器人

12.6 Welding robot production line 焊接机器人生产线

The simple welding robot production line is a production line which connects workstations with conveying lines. This production line maintains the single-station character, with each station performing scheduled welding work using only selected fixtures and robotic programs.

简单的焊接机器人生产线是把多台工作站用工件输送线连接起来组成的生产线。这种生产线保持单站的特点，每个站只能使用选定的夹具及机器人程序执行预定的焊接工作。

The welding flexible production line is composed of multiple stations, where workpiece are clamped on a unified form of tray that can be matched with a transposition machine at any station on the production line. The welding robot first identifies the serial number of the tray or the workpiece and automatically calls out the program of welding this kind of workpiece. Fig.12-9 shows the welding robot on the production line.

焊接柔性生产线由多个站组成，焊件都装夹在统一形式的托盘上，托盘可以与生产线上任何一个站的变位机相配合。焊接机器人首先对托盘的编号或工件进行识别，自动调出焊接这种工件的程序。图 12-9 所示是生产线上的焊接机器人。

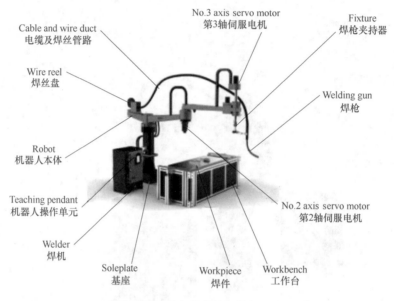

图 12-9　生产线上的焊接机器人

Each station can weld different workpieces without any adjustments. The welding flexible line usually has a track car, which can automatically remove the workpiece from the storage station and send it to the machine with space. The welded workpiece can also be removed from the workstation and sent to the finished workpiece location.

每一个站无需作任何调整就可以焊接不同的工件。焊接柔性线一般有一个轨道小车，小车可以自动将工件从存放工位取出，再送到有空位的变位机上。也可以从工作站上把焊接完毕的工件取下，送到成品工件位置。

The whole flexible welding production line is controlled by a computer. Therefore, as long as enough of the workpiece is assembled during the day and put on the storage station, the production can be realized at night with no or few people.

整条柔性焊接生产线由一台计算机控制。因此，只要白天装配好足够多的工件，并放到存放工位上，夜间就可以实现无人或少人生产了。

Welding machine is suitable for large batch, slow modification products, suitable for the workpiece of the weld less, longer, shape rules（straight, round); welding robot system is generally suitable for medium and small batch production, the welds can be short and many, the shape is more complex. Flexible welding line is especially suitable for the product variety, each batch of small quantities.

焊接专机适合批量大、改型慢的产品，适用于工件的焊缝数量较少、较长、形状规矩（直线、圆形）的情况；焊接机器人系统一般适合中、小批量生产，被焊工件的焊缝可以短而多，形状较复杂。柔性焊接线特别适合产品品种多、每批数量又很少的情况。

12.7 Application of welding robot in automobile production
焊接机器人在汽车生产中的应用

At present, welding robot has been widely used in automobile manufacturing industry, which is used in the welding of automobile chassis, seat frame, guide rail, muffler, hydraulic torque converter and other parts, especially in the automobile chassis welding production has been widely used.

焊接机器人目前已广泛应用在汽车制造业，用于汽车底盘、座椅骨架、导轨、消声器以及液力变矩器等部件的焊接，尤其在汽车底盘焊接生产中得到了广泛的应用。

Domestic production of Santana, Passat, Buick and other rear axle, auxiliary frame, rocker arm, suspension, shock absorber and other car chassis parts mostly use MIG welding process production, the main components adopt stamping welding, plate thickness is 1.5~4mm on average, welding mainly in the form of lap joint, angle joint.

国内生产的桑塔纳、帕萨特、别克等轿车的后桥、副车架、摇臂、悬架、减振器等底盘零件多使用 MIG 焊接工艺生产，主要构件采用冲压焊接，板厚平均为 1.5~4mm，焊接主要以搭接、角接接头形式为主。

Welding quality requirements are quite high, the quality of welding directly affects the safety performance of cars. After the application of robot welding, the appearance and internal quality of welds are greatly improved, and the quality stability is guaranteed, labor intensity is reduced, and the working environment is improved. Fig.12-10 shows the application of robots in the automotive industry.

焊接质量要求相当高，焊接质量的好坏直接影响到轿车的安全性能。应用机器人焊接后，大大提高了焊接件的外观和内在质量，并保证了质量的稳定性，降低了劳动强度，改善了劳动环境。图 12-10 所示是机器人在汽车行业中的应用。

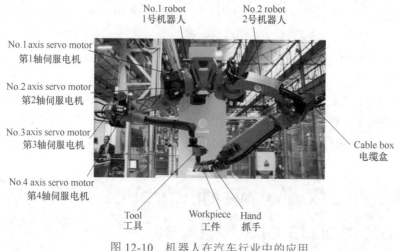

图 12-10 机器人在汽车行业中的应用

12.8 Characteristics of arc welding　弧焊特点

12.8.1　Basic function　基本功能

Arc welding process is much more complex than spot welding process, tool center point (TCP), that is, the movement trajectory of the wire end, the attitude of the welding gun, welding parameters are required to be precisely controlled. Therefore, in addition to the general functions mentioned above, arc welding robot must also have some functions suitable for arc welding requirements.

弧焊过程比点焊过程复杂得多,工具中心点(TCP),即焊丝端头的运动轨迹、焊枪姿态、焊接参数,都要求精确控制。所以,弧焊用机器人除了前面所述的一般功能外,还必须具备一些适合弧焊要求的功能。

In theory, a robot with five axes could be used for arc welding, but for welds of complex shapes, a robot with five axes would have difficulty. Therefore, unless the weld is relatively simple, otherwise should try to choose the 6-axis robot.

从理论上讲,有5个轴的机器人就可以用于电弧焊,但是对复杂形状的焊缝,用5轴机器人会有困难。因此,除非焊缝比较简单,否则应尽量选用6轴机器人。

Arc welding robot for welding the corner or small diameter round welding, its trajectory, should be able to close to the teaching should also have different styles of software function, for programming, in order to make swing welding, and the pause in each cycle of bead welding points, robots should automatically stop moving forward, in order to meet the technological requirements. In addition, there should be contact location, automatic finding of the starting position of the weld, arc tracking and automatic restarting functions. Fig. 12-11 shows the arc welding robot.

弧焊机器人作拐角焊或小直径圆焊接时,其轨迹应能贴近示教的轨迹,还应具备不同摆动样式的软件功能,供编程时选用,以便作摆动焊,而且在摆动焊接的每一周期中的停顿点处，机器人应自动停止向前运动，以满足工艺要求。此外，还应有接触寻位、自动寻找焊缝起点位置、电弧跟踪及自动再引弧功能等。图 12-11 所示是弧焊机器人。

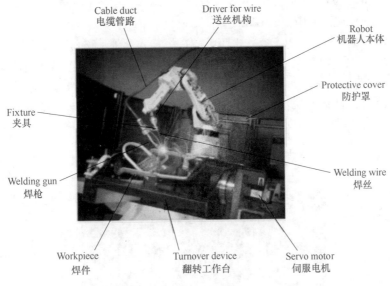

图 12-11　弧焊机器人

12.8.2　Welding equipment　焊接设备

　　Arc welding robot mostly uses gas shielded welding method (MAG, MIG, TIG), usual thyristor, inverter, waveform control, pulse or non-pulse welding power can be installed on the robot for arc welding.

　　弧焊机器人多采用气体保护焊方法（MAG、MIG、TIG），通常的晶闸管式、逆变式、波形控制式、脉冲或非脉冲式等的焊接电源都可以装到机器人上作电弧焊。

　　Since the robot controller adopts digital control, while the welding power supply is mostly analog control, an interface should be added between the welding power supply and the controller. Robot manufacturers have their own specific welding equipment, these welding equipment has been configured with the corresponding interface board. Arc time accounts for a large proportion in the working cycle of arc welding robot, so when choosing welding power supply, the capacity of power supply should be determined according to the duration rate of 100%.

　　由于机器人控制器采用数字控制，而焊接电源多为模拟控制，所以需要在焊接电源与控制器之间加一个接口。机器人生产厂都有自己特定的配套焊接设备，这些焊接设备已经配置相应的接口板。在弧焊机器人工作周期中起弧时间所占的比例较大，因此在选择焊接电源时，一般应按持续率 100% 来确定电源的容量。

　　The wire feeder can be mounted on the upper arm of the robot or placed outside the robot. The wire feeder is mounted on the upper arm of the robot. The hose between the welding gun and the wire feeder is shorter, which is conducive to maintaining the stability of wire feeder. The wire feeding device is installed outside the robot, and the hose is a little long. When the robot sends the welding torch to some positions, the hose will be in a state of multiple bending, which will seriously affect the quality of wire feeding. So the installation of wire feeding machine must ensure the stability of wire feeding. Fig. 12-12 shows the engineer adjusting the running trajectory of the welding robot.

送丝装置可以装在机器人的上臂，也可以放在机器人之外。送丝装置装在机器人的上臂，焊枪到送丝机之间的软管较短，有利于保持送丝的稳定性。送丝装置装在机器人之外，软管较长，当机器人把焊枪送到某些位置，使软管处于多弯曲状态，会严重影响送丝的质量。所以送丝机的安装方式必须要保证送丝稳定性。图12-12所示是工程师调整焊接机器人运行轨迹。

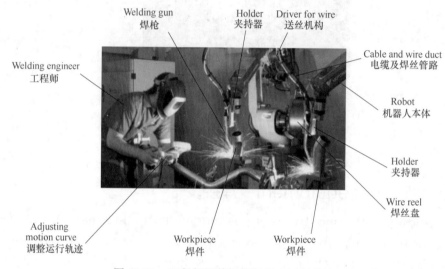

图 12-12　工程师调整焊接机器人运行轨迹

12.8.3　Maintaining　维护保养

Daily inspection and maintenance.
每日检查及维护。

1）Wire feeding mechanism. Including whether the wire feeding distance is normal, whether the wire feeding catheter is damaged, whether there is an abnormal alarm.

送丝机构。包括送丝距离是否正常，送丝导管是否损坏，有无异常报警。

2）Whether the gas flow is normal.
气体流量是否正常。

3）Whether the welding gun safety protection system is normal.
焊枪安全保护系统是否正常。

4）Whether the water circulation system works properly.
水循环系统工作是否正常。

Chapter 13
The Application of Robots in Other Industries
机器人在其他行业中的应用

13.1 Application of robot in transportation industry
机器人在搬运码垛行业中的应用

The handling robot is an industrial robot that can carry out automatic handling. Handling operation refers to the movement of a workpiece from one processing position to another with a device. Handling robot can be installed with different hand to complete a variety of different workpiece handling, greatly reducing the heavy manual labor of human.

搬运机器人是可以进行自动搬运作业的工业机器人。搬运作业是指用一种设备夹持工件从一个加工位置移到另一个加工位置。搬运机器人可安装不同的抓手以完成各种不同工件搬运工作,大大减轻了人类繁重的体力劳动。

Handling robot is widely used in machine tool loading and unloading, punch automatic production line, automatic assembly line, stack handling, container transportation and other automatic handling. Some developed countries have set the maximum limit for manual handling, which must be completed by handling robots. Fig. 13-1 shows the composition of the transport robot.

搬运机器人被广泛应用于机床上下料、冲压机自动化生产线、自动装配流水线、码垛搬运、集装箱运输等自动搬运工作。部分发达国家已制定出人工搬运的最大限度,超过限度的工作必须由搬运机器人完成。图 13-1 所示为搬运机器人的构成。

Handling robot is a high and new technology, involving mechanics, electrical hydraulic pressure technology, automatic control technology, sensor technology, single-chip microcomputer technology and computer technology and other fields, has become an important part of modern machinery manufacturing production system. The advantage of the transport robot is that it can be programmed to complete a variety of expected tasks, combining the advantages of man and machine in its own structure and performance, especially artificial intelligence and adaptability. Fig. 13-2 shows the handling robot applied to the filling production line.

搬运机器人是一项高新技术，涉及到力学、机械学、电气液压气压技术、自动控制技术、传感器技术、单片机技术和计算机技术等学科领域，已成为现代机械制造生产体系中的一项重要组成部分。搬运机器人的优点是可以通过编程完成各种预期的任务，在自身结构和性能上结合了人和机器的各自优势，尤其是人工智能和适应性。图 13-2 所示是搬运机器人应用于灌装生产线。

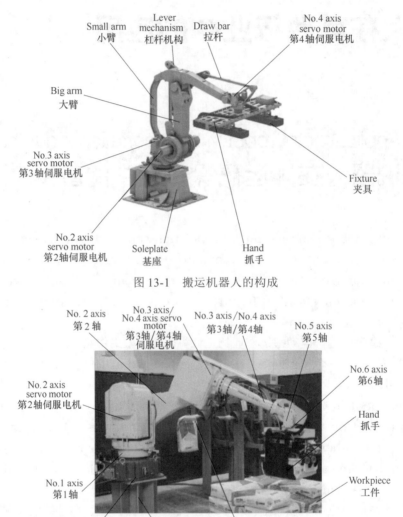

图 13-1 搬运机器人的构成

图 13-2 搬运机器人应用于灌装生产线

When using palletizing robots, one important thing is how the robot grabs a product. Vacuum gripper is the most common gripper. Vacuum grippers are cheap, easy to operate, and can load most items efficiently. However, in some specific

Chapter 13 The Application of Robots in Other Industries

applications, the vacuum gripper can also encounter problems, such as the surface of the porous workpiece, the contents of the liquid packaging, or the surface of the uneven packaging and so on.

在使用码垛机器人的时候,一个重要的事情,就是机器人怎样抓住一个产品。真空抓手是最常见的抓手。真空抓手价格便宜,易于操作,而且能够有效装载大部分物品。但是在一些特定的应用中,真空抓手也会遇到问题,例如表面多孔的工件,内容物为液体的软包装,或者表面不平整的包装等等。

Other gripper types include the clamshell gripper, which holds the sides of a bag or other item. The fork gripper inserts into the bottom of the item to lift it up. There's also the bag gripper, a cross between the clamshell and fork gripper, whose fork portion wraps around the bottom and sides of the item. Fig. 13-3 shows the handling robot grabbing objects.

其他的抓手类型包括翻盖式抓手,它能将一个袋子或者其他物品的两边夹住;叉式抓手,它插入物品的底部将物品提升起来;还有袋式抓手,这是翻盖式和叉子式抓手的混合体,它的叉子部分能包裹住物品的底部和两边。图 13-3 所示是搬运机器人抓取物品。

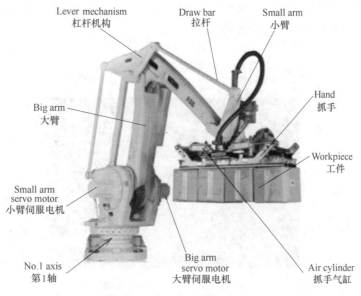

图 13-3 搬运机器人抓取物品

Feature 特点:

1) Simple structure and few parts. Therefore, the failure rate of the parts is low, the performance is reliable, and the maintenance is simple.

结构简单、零部件少,因此零部件的故障率低、性能可靠、保养维修简单。

2) It takes up less space. It is beneficial to the arrangement of production line in the customer factory. Palletizing robots can be set up in narrow spaces for efficient use.

占地面积少。有利于客户厂房中生产线的布置。码垛机器人可以设置在狭窄的空间、有效地使用。

3) Excellent applicability. When the product size, volume, shape and pallet size change, just modify the parameters on the touch screen, which will not affect the normal production of the customer.

优良的适用性。当产品的尺寸、体积、形状及托盘的外形尺寸发生变化时只需在触摸屏上做修改参数即可，不会影响客户正常的生产。

4) Low energy consumption. Generally, the power of mechanical palletizer is about 26kW, while the power of palletizer is about 5kW. Greatly reduced the customer's operating costs.

能耗低。通常机械式的码垛机的功率在 26kW 左右，而码垛机器人的功率为 5kW 左右，大大降低了客户的运行成本。

5) All controls can be operated on the GOT, simply.

全部控制可在控制柜屏幕上操作，操作简单。

6) Just position the starting point and put the point, the teaching method is simple and easy to understand.

只需定位起点和摆放点，示教方法简单易懂。

Fig. 13-4 shows the application of the handling robot on the transportation line.

图 13-4 所示是搬运机器人在运输线上的应用。

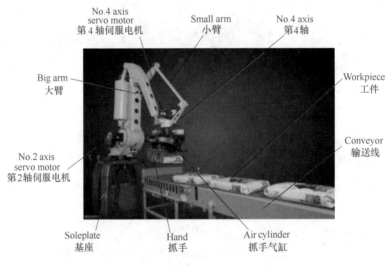

图 13-4 搬运机器人在运输线上的应用

13.2 Application of robot in polishing industry 机器人在打磨抛光行业中的应用

（1）Overview 概述

Polishing robot is a kind of industrial robot. It is used to replace the manual polishing work of the workpiece. Polishing robot is mainly used for high-precision

Chapter 13 The Application of Robots in Other Industries

sanding and polishing of sanitary ware, auto parts, industrial parts, medical devices, civil products and other industries. It is used for surface polishing, edge and angle deburring, welding seam polishing, inner hole deburring and other work. Fig. 13-5 shows the polishing robot processing the bathroom parts.

打磨抛光机器人是工业机器人的一种。用于替代人工进行工件的打磨抛光工作。打磨机器人主要用于卫浴行业、汽车零部件、工业零件、医疗器械、民用产品等行业高精度的打磨抛光作业。它被用于工件的表面打磨、棱角去毛刺、焊缝打磨、内孔去毛刺等工作。

图 13-5 所示是抛光机器人在加工卫浴零件。

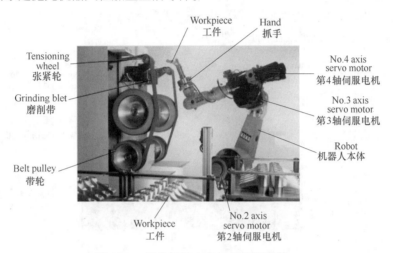

图 13-5 抛光机器人在加工卫浴零件

Manual deburring has the following disadvantages: time-consuming, poor polishing effect, low efficiency, and the operator's hand is often injured. The air pollution and noise in the polishing work site will harm the operator's physical and mental health. The polishing robot is equipped with different grinding machines and grinding heads according to the requirements of the finished parts. The polishing robot can carry out polishing work for a long time and has high productivity. Ensure product high quality and high stability. Fig. 13-6 shows the polishing robot deburring.

人工去毛刺有以下缺点：费时、打磨效果不好、效率低，而且操作者的手还常常受伤。打磨工作现场的空气污染和噪声会损害操作者的身心健康。打磨机器人根据被加工零部件光洁度要求配置不同的打磨机和磨头。打磨机器人可长期进行打磨作业、具有高生产率，可保证产品高质量和高稳定性。图 13-6 所示是抛光机器人的去毛刺加工。

（2）The main advantages of robot polishing 机器人打磨的主要优点

Improve polishing quality and product finish to ensure consistency.

Increase productivity, 24 hours a day continuous production.

Improve the working conditions of workers, can work in harmful environment for a long time.

Reduce the requirement for workers' operation skills.

Shorten product modification cycle, reduce investment in equipment.

With redevelopable, users can be based on the same piece of secondary programming.

Fig. 13-7 shows the polishing robot polishing the surface of workpiece.

提高打磨质量和产品光洁度，保证产品质量一致性。

提高生产率，可 24h 连续生产。

改善工人劳动条件，可在有害环境下长期工作。

降低对工人操作技术的要求。

缩短产品改型的周期，减少投资设备。

具备可再开发性，用户可根据不同样件进行二次编程。

图 13-7 所示是抛光机器人在抛光工件表面。

图 13-6　抛光机器人在做去毛刺加工

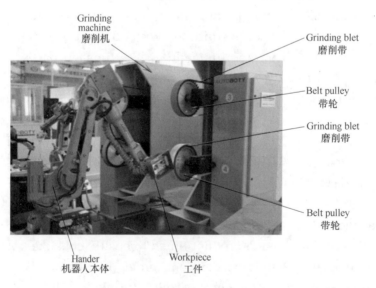

图 13-7　抛光机器人在抛光工件表面

(3) Polishing robot composition　打磨机器人的组成

Polishing robot is generally composed of teaching unit, control cabinet, robot

Chapter 13 The Application of Robots in Other Industries

body, pressure sensor, grinding machinery and other parts. Continuous trajectory control and point control can be realized. Fig. 13-8 and Fig. 13-9 show the robot polishing the surface of the workpiece.

打磨机器人一般由示教单元、控制柜、机器人本体、压力传感器、磨削机械等部分组成。可以实现连续轨迹控制和点位控制。图 13-8 和图 13-9 所示是机器人在打磨抛光工件表面。

图 13-8　机器人在打磨工件表面

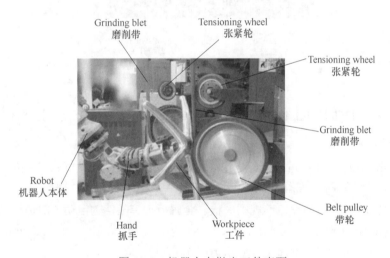

图 13-9　机器人在抛光工件表面

13.3　Application of robot in machine tool industry
机器人在机床加工行业中的应用

13.3.1　Overview　概述

Loading and unloading robot can meet the "quick/bulk processing", "save the human cost", "efficiency" requirements, such as loading robot system with high efficiency and high stability, simple in structure, easy to maintain, can satisfy the production of different products, can quickly adjust product structure and capacity expansion, to reduce the labor intensity of industry workers. Fig. 13-10 shows the application of the loading and unloading robot in machine tool processing.

上下料机器人能满足"快速/大批量加工""节省人力成本""提高生产效率"等要求，上下料机器人系统具有高效率和高稳定性，结构简单易于维护，可以满足不同种类产品的生产，可以快速进行产品结构的调整和扩大产能，降低产业工人的劳动强度。图 13-10 是上下料机器人在机床加工中的应用。

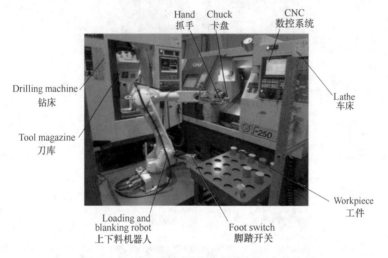

图 13-10　上下料机器人在机床加工中的应用

13.3.2　Loading and unloading robot features　上下料机器人的特点

1) It can realize automatic loading/unloading for disc class, long-axis class, irregular shape and metal plate class.

可以实现对圆盘类、长轴类、不规则形状、金属板类等工件的自动上料/下料。

2) Robot with independent control module, combined with the control system of the machine to control, do not affect the machine operation.

机械手采用独立的控制模块，结合机床的控制系统进行控制，不影响机床运转。

3) Good rigidity, smooth operation, easy maintenance.

刚性好，运行平稳，维护非常方便。

4) Optional: separate bin design, separate bin automatic control.

可选：独立料仓设计，料仓独立自动控制。

5) Optional: separate assembly line.

可选：独立流水线。

Fig. 13-11 shows the robot loading and unloading the CNC lathe.

图 13-11 所示是机器人给数控车床上下料。

Chapter 13 The Application of Robots in Other Industries

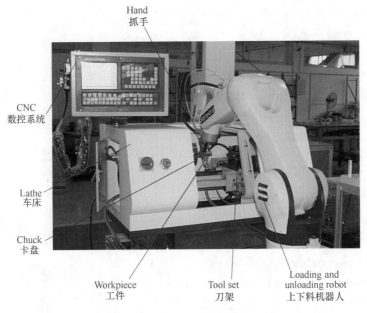

图 13-11 机器人给数控车床上下料

13.3.3 Robot classification 机器人分类

（1）**Prosthetic robot** 关节式机器人（某品牌的技术指标）

1）Maximum working range of robot (radius of rotation): 620~3503mm.

机器人最大工作范围（回转半径）：620~3503mm。

2）Robot load capacity: 3~700kg.

机器人负载能力：3~700kg。

3）Robot working tempo: ≥3s.

机器人工作节拍：≥3s。

4）Positioning accuracy: ±0.1mm.

定位精度：±0.1mm。

5）Drive type: full servo drive.

驱动形式：全伺服驱动。

6）Gripper drive: pneumatic or electric, with automatic replacement gripper function.

抓手驱动：气动或者电动，具备自动更换抓手功能。

Fig. 13-12 shows the robot loading and unloading two CNC lathes.

图 13-12 所示是机器人给两台数控车床上下料。

（2）**Rectangular coordinate robot** 直角坐标机器人

1）Robot working range 机器人工作范围：

a）Horizontal stroke: 1000~20000mm.

水平行程：1000~20000mm。

b）Vertical travel: 200~3000mm.

垂直行程：200~3000mm。

c）Workpiece rotation：±180°.

工件旋转：±180°。

2）Speed　运行速度：

Maximum speed of horizontal movement：3000mm/s.

水平运动最大速度：3000mm/s。

Maximum vertical motion speed：1000mm/s.

垂直运动最大速度：1000mm/s。

3）Positioning accuracy　定位精度：

Horizontal motion repetition accuracy：±0.1mm.

水平运动重复精度：±0.1mm。

Vertical motion repetition accuracy：±0.1mm.

垂直运动重复精度：±0.1mm。

4）Type of belt drive　传动形式：

a）Horizontal motion transmission：synchronous belt/rack and pinion.

水平运动传动形式：同步带/齿轮齿条。

b）Vertical motion transmission：synchronous belt/rack and pinion/screw.

垂直运动传动形式：同步带/齿轮齿条/丝杆。

5）Load weight：the maximum load is 1000kg.

负载质量：最大负载1000kg。

6）Motion control system：PLC/ motion control card /CNC.

运动控制系统：PLC/运动控制卡/CNC。

7）Gripper drive：pneumatic or electric, with automatic replacement gripper function.

抓手驱动：气动或者电动，具备自动更换抓手功能。

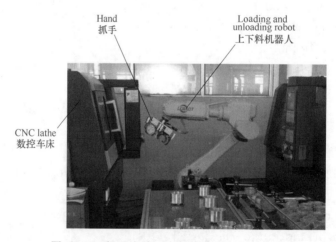

图 13-12　机器人给两台数控车床上下料

　　Both joint robot and coordinate robot can do loading and unloading work well. Articular robots work efficiently, move quickly, occupy less space, but cost more. The coordinate robot is efficient and occupies a large space, but the cost is low.

Chapter 13　The Application of Robots in Other Industries

关节型机器人和坐标型机器人都可以很好地完成上下料工作。关节型机器人工作效率高，动作节拍快，占地空间小，但是成本高。坐标型机器人工作效率高，占地空间大，但成本低。

The loading and unloading robot system mainly consists of robot, silo system, gripper system, control system and safety protection system. The production line is composed of CNC machine tool system. Fig. 13-13 shows the robot feeding the flanging machine.

上下料机器人系统主要由机器人、料仓系统、抓手夹持系统、控制系统、安全防护系统组成，与数控机床系统组成生产线。图 13-13 所示是机器人给折边机床上下料。

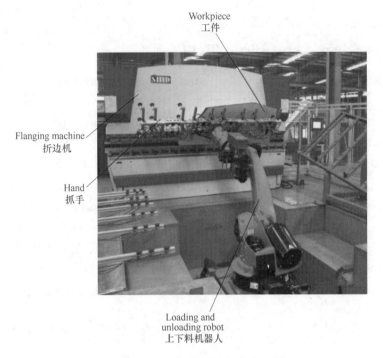

图 13-13　机器人给折边机床上下料

The gripper of the robot system adopts pneumatic mechanical structure and has two types: double gripper and single gripper. It is safe, firm and reliable, with no loss of parts, and can realize power failure and cut off workpiece clamping. Gripper clamps workpiece by friction force and clamping force, at the same time uses locating pin for auxiliary positioning. When the clamping surface is blank surface, the clamping block floats, in order to adapt to the casting deviation of workpiece shape.

机器人系统的抓手采用气动机械结构，有双抓手和单抓手两种类型。安全牢固可靠，无掉件现象，能够实现断电、断气工件夹紧。抓手夹紧工件时，靠摩擦力及夹紧力夹紧，同时用定位销辅助定位，当夹紧面为毛坯面时，夹紧块浮动，适应工件外形的铸造偏差。

Machine tool feeding and unloading robot has the following characteristics:
机床上下料机器人具有以下特点：

1) Programmable. The robot can be reprogrammed as the job demands change. Therefore, it is an important part of the flexible manufacturing system.

可编程。机器人可以随工作要求的变化再编程，因此它是柔性制造系统的重要组成部分。

2) Personified. Intelligent robot can be equipped with contact sensor, force sensor, load sensor, which improved the robot's adaptive ability.

拟人化。智能化机器人可以装备接触传感器、力觉传感器、负载传感器，提高了机器人的自适应能力。

3) Universality. After the robot's gripper or tool is replaced, different jobs can be performed.

通用性。在更换机器人的抓手或工具后，可以执行不同的工作。

13.3.4　Working example　工作样例

To improve production efficiency, some robot systems are equipped with two grippers. Two workpieces can be clamped at the same time. The robot first picks up a blank, when the machine tool finished processing a workpiece, open the door, the robot hand goes into the machine tool, clamp the processed workpiece, after rotating 180°, and clamps blank. Then exits the machine, puts the workpiece, and reclamps the blank for processing. It saves the huge workpiece conveying device. Fig. 13-14 shows the robot loading, unloading and clamping the CNC lathe.

为了提高生产效率，有些机器人系统装有 2 个抓手，可以同时夹住 2 个工件。机器人首先夹起一个毛坯，当机床加工完毕一个工件，打开门后，机械手进入机床内，夹取加工完毕的工件，旋转 180° 后，夹装毛坯。然后退出机床、摆放工件、重新夹装毛坯等待加工。这样节省了庞大的工件输送装置。图 13-14 所示是机器人给数控车床上下料并调头装夹。

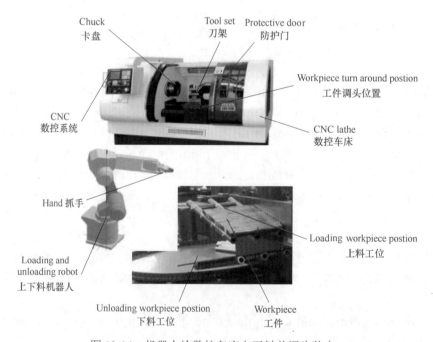

图 13-14　机器人给数控车床上下料并调头装夹

Chapter 13 The Application of Robots in Other Industries

（1）The general flow chart of robot unloading and clamping 机器人卸料装夹总流程图（图 13-15）

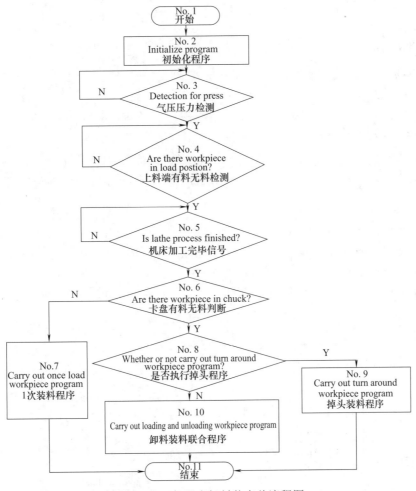

图 13-15 机器人卸料装夹总流程图

（2）Main program 卸料装夹总程序

```
1   CALLP"CHUSH"                    '——调用初始化程序。
2   *LAB3                           '——程序分支标志。
3   *YALI                           '——程序分支标志。
4   IF M15=0 THEN GOTO *YALI        '——判断气压是否达到标准。
5   *QULIAO                         '——程序分支标志。
6   IF M25=0 THEN GOTO *QULIAO      '——判断上料端有料无料。
7   *WANC                           '——程序分支标志。
8   IF M35=0 THEN GOTO *WANC        '——判断机床加工是否完成信号。
9   IF M100=0 THEN GOTO *LAB1       '——判断是否执行1次上料。
10  IF M200=0 THEN GOTO *LAB2       '——判断是否执行调头装夹。
11  CALLP"XANDJ"                    '——调用卸料装夹联合程序。
```

```
12    M300=1                    '——卸料装夹联合程序执行完毕。
13    M200=0                    '——可执行掉头装夹。
14    END                       '——主程序结束。
15    *LAB1                     '——执行首次上料。
16    CALLP"FIRST"              '——调用首次上料程序。
17    M100=1                    '——首次上料执行完毕。
18    GOTO*LAB3                 '——跳转到 *LAB3 行。
19    *LAB2                     '——执行调头装夹程序。
20    CALLP "EXC"               '——调用调头装夹程序。
21    M200=1                    '——掉头装夹执行完毕。
22    GOTO*LAB3                 '——跳转到 *LAB3 行。
```

(3) Once clamping 一次装夹

Once clamping refers to in the machine tool processing, with no workpiece on the chuck, only perform once clamping workpiece. Fig. 13-16 shows the flow chart of the once clamping.

一次装夹指在机床加工时，卡盘上无工件，只执行一次装夹工件。图 13-16 所示是一次装夹流程图。

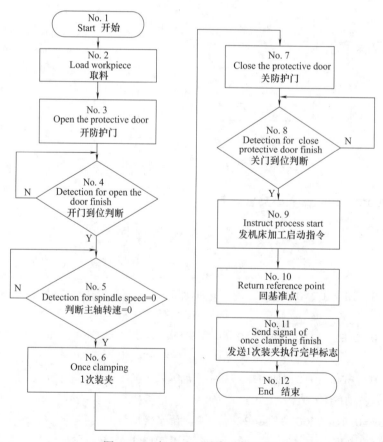

图 13-16 机器人一次装夹流程图

Chapter 13   The Application of Robots in Other Industries

Fig. 13-17 shows the motion path of the robot to perform once clamping.
图 13-17 所示是机器人执行一次装夹的运动路径。

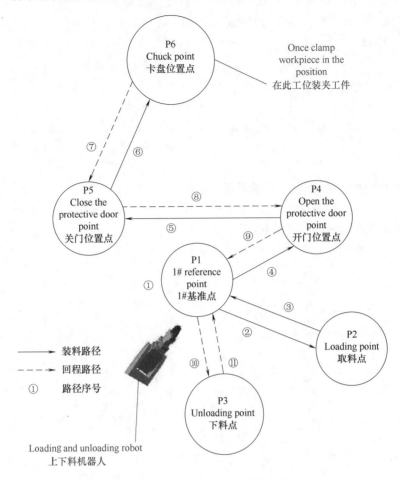

图 13-17   机器人执行一次装夹的路径

As shown in Fig. 13-17, the path for the robot to perform once clamping operation is as follows.

如图 13-17，机器人执行一次装夹的路径如下所示。

P1 → P2 → P1 → P4 → P5 → P6 → P5 → P4 → P1 → P3 → P1

P1——1# reference point/1# 基准点。

P2——loading point/ 取料点。

P3——unloading point/ 下料点。

P4——open the protective door point/ 开门位置点。

P5——close the protective door point/ 关门位置点。

P6——chuck point/ 卡盘位置点。

（4）Once clamping program   一次装夹程序

```
1  CallP "QUL"                       '——调用取料子程序。
2  CallP "KAIM"                      '——调用开门子程序。
```

```
3   *LAB1                                '——程序分支标志。
4   IF M_IN(11)=1 THEN GOTO*LAB1         '——主轴速度=0判断。
5   '——如果主轴速度不为0，则跳转到*LAB1，否则执行下一步。
6   CallP "JIAZ"                         '——调用夹装子程序。
7   CallP "GM"                           '——调用关门子程序。
8   M_OUT(17)=1                          '——发机床加工启动指令。
9   MOV P1                               '——回基准点。
10  M100=1                               '——发首次装夹完成标志。
11  END                                  '——程序结束。
```

（5）Turning and clamping　调头装夹

Fig. 13-18 is the flow chart of the robot turning and clamping program.
图 13-18 是机器人调头装夹程序流程图。

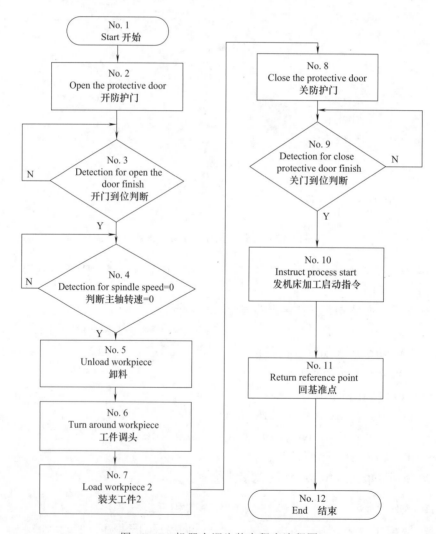

图 13-18　机器人调头装夹程序流程图

Chapter 13   The Application of Robots in Other Industries

Fig. 13-19 shows the movement path of the robot to perform the turning and clamping.

图 13-19 所示是机器人执行调头装夹的运动路径。

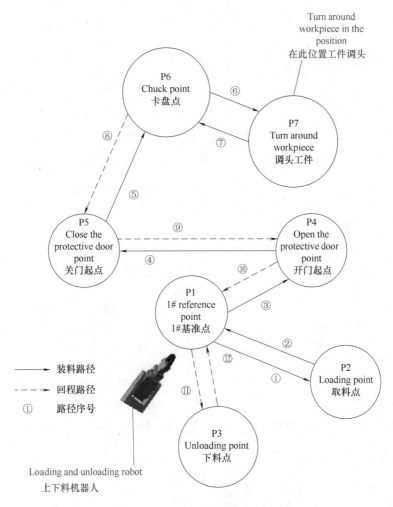

图 13-19   机器人执行调头装夹的路径

As shown in Fig. 13-19, the robot performs the turning and the clamping path is as follows.

如图 13-19，机器人执行调头装夹的路径如下所示。

　　P1 → P2 → P1 → P4 → P5 → P6 → P7 → P6 → P5 → P4 → P1 → P3 → P1

P1——1# reference point/1# 基准点。
P2——loading point/ 取料点。
P3——unloading point/ 下料点。
P4——open the protective door point/ 开门位置点。
P5——close the protective door point/ 关门位置点。
P6——chuck point/ 卡盘位置点。
P7——turn around workpiece/ 调头工件。

（6）Unloading and clamping　卸料装夹

Fig. 13-20 is the flow chart of the program of robot unloading and clamping.
图 13-20 是机器人卸料装夹联合程序流程图。

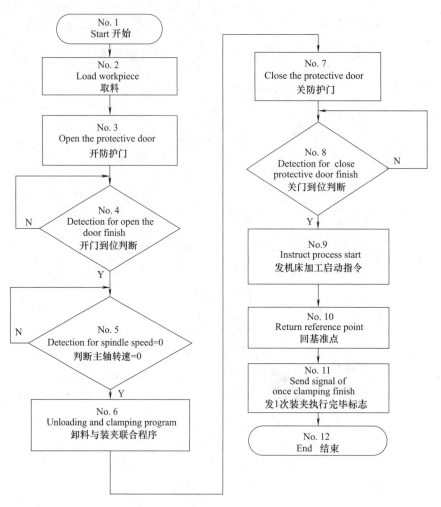

图 13-20　卸料装夹联合程序图

（7）Program of unloading and clamping　卸料装夹程序

```
1   CallP "KAIM"                        '——调用开门子程序。
2   *LAB1                               '——程序分支标志。
3   IF M_IN（11）=1 THEN GOTO*LAB1      '——主轴速度=0 判断。
4   '——如果主轴速度不为 0，则跳转到 *LAB1。
5   CallP "XJ"                          '——调用卸料装夹子程序。
6   CallP "GM"                          '——调用关门子程序。
7   M_OUT（17）=1                       '——发机床加工启动指令。
8   CallP "xial"                        '——调用下料子程序。
9   END                                 '——程序结束。
```

Chapter 13　The Application of Robots in Other Industries

13.4　Application of robot in cutting industry
机器人在切割行业中的应用

Fig. 13-21 is the application of robot in gantry cutting machine.
图 13-21 是机器人在龙门切割机中的应用。

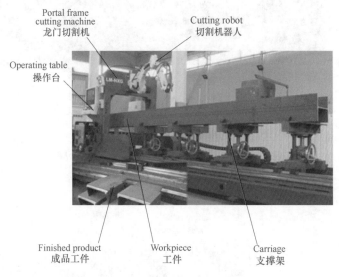

图 13-21　机器人在龙门切割机中的应用

Fig. 13-22 shows the robot cutting steel pipe.
图 13-22 是机器人切割钢管类部件。

图 13-22　机器人切割钢管类部件

Fig. 13-23 shows the cutting torch attached to the robot.
图 13-23 所示是连接在机器人上的割炬。

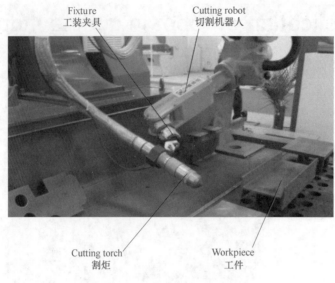

图 13-23　连接在机器人上的割炬

Fig. 13-24 shows the robot cutting steel plate.

图 13-24 所示是机器人切割钢板。

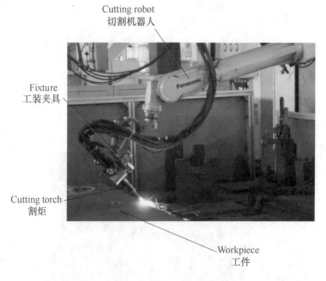

图 13-24　机器人切割钢板

13.5　Application of robot in spraying industry 机器人在喷涂行业中的应用

Spraying robot is an industrial robot that can spray paint or other coatings automatically. Paint robot is composed of robot body, computer and corresponding

Chapter 13 The Application of Robots in Other Industries

control system. With 5 or 6 degrees of freedom joint structure, the arm has a large movement space, and can do complex trajectory movement, the wrist generally has 2 to 3 degrees of freedom, with flexible movement. Fig. 13-25 is the spraying robot in the spraying operation.

喷涂机器人是可进行自动喷漆或喷涂其他涂料的工业机器人。喷漆机器人由机器人本体、计算机和相应的控制系统组成。采用 5 或 6 自由度关节式结构，手臂有较大的运动空间，并可做复杂的轨迹运动，其腕部一般有 2~3 个自由度，可灵活运动。图 13-25 是喷涂机器人在喷涂作业。

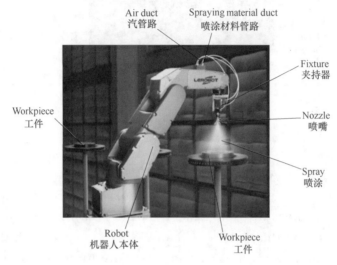

图 13-25　喷涂机器人在喷涂作业

The advanced spray-painting robot adopts flexible wrist, which can be bent in all directions and can be rotated. Its action is similar to that of human wrist, which can be easily inserted into the workpiece through small holes to spray its inner surface. Spray-painting robots are widely used in automotive, instrumentation, electrical appliances, enamel and other process production departments. Fig. 13-26 shows the robot spraying an automobile part.

先进的喷漆机器人采用柔性手腕，既可向各个方向弯曲，又可转动，其动作类似人的手腕，能方便地通过较小的孔伸入工件内部，喷涂其内表面。喷漆机器人广泛用于汽车、仪表、电器、搪瓷等工艺生产部门。图 13-26 是机器人喷涂汽车部件。

The main advantages of spraying robot are good flexibility and large scope of work. It can improve spraying quality and material utilization, and is easy to operate and maintain. Spraying robot can be programmed off-line to shorten the time of field debugging. The equipment utilization rate of spraying robot can reach 90%~95%.

喷涂机器人的主要优点是柔性好、工作范围大。能提高喷涂质量和材料使用率、易于操作和维护。喷涂机器人可离线编程，缩短现场调试时间。喷涂机器人的设备利用率可达 90%~95%。

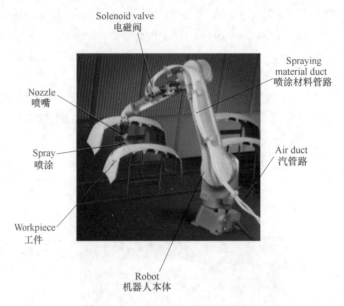

图 13-26 机器人喷涂汽车部件

Classify 分类：

(1) Air spraying robot 有气喷涂机器人

Air spraying robot is also called low pressure air spraying robot. The spray machine relies on low pressure air to make the paint spray nozzle after the formation of atomized air acting on the surface of the object. Compared with the hand brush, there is no brush mark, and the plane is relatively uniform, unit working time is short, and it can effectively shorten the period. But there is air spray splash phenomenon, resulting in paint waste.

有气喷涂机器人也称低压有气喷涂机器人。喷涂机依靠低压空气使油漆在喷出枪口后形成雾化气流作用于物体表面。有气喷涂相对于手刷而言无刷痕，而且平面相对均匀，单位工作时间短，可有效地缩短工期。但有气喷涂有飞溅现象，造成漆料浪费。

(2) Airless spray robot 无气喷涂机器人

Airless spraying robot can be used in the construction of high viscosity paint. It can be divided into pneumatic airless spraying machine, electric airless spraying machine, internal combustion airless spraying machine, automatic spraying machine and so on.

Fig. 13-27 shows the spraying robot working in the spraying workshop.

无气喷涂机器人可用于高黏度油漆的施工，可分为气动式无气喷涂机、电动式无气喷涂机、内燃式无气喷涂机、自动喷涂机等多种。图 13-27 所示是喷涂机器人在喷涂工作间内作业。

Selection technical parameters are as follows.

选型技术参数如下。

1) The working range of the robot. When selecting the robot, it is necessary to ensure that the working trajectory of the robot can completely cover the relevant

Chapter 13　The Application of Robots in Other Industries

surface or internal cavity of the workpiece to be constructed.

机器人的工作轨迹范围。在选择机器人时需保证机器人的工作轨迹范围必须能够完全覆盖所需施工的工件的相关表面或内腔。

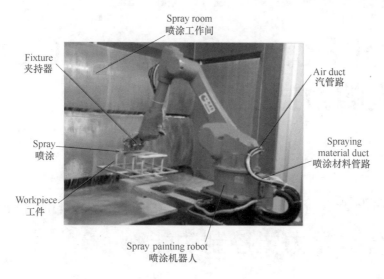

图 13-27　喷涂机器人在喷涂工作间内作业

2）Robot repetition accuracy. The repeat accuracy of the gluing robot only needs to reach 0.5mm. For paint robots, the repetition accuracy can be lower.

机器人的重复精度。涂胶机器人的重复精度只需达到 0.5mm。对于喷漆机器人，重复精度可更低。

3）Largest load. Different spraying work with different weight of the spray, the maximum carrying capacity of the robot to meet the requirements.

最大荷载。不同的喷涂工作配置不同重量的喷具，要求机器人的最大承载能力满足要求。

Fig. 13-28 shows the spraying robot spraying the car shell.

图 13-28 所示是喷涂机器人对汽车外壳进行喷涂作业。

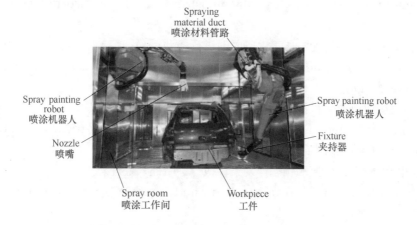

图 13-28　喷涂机器人对汽车外壳进行喷涂作业

Chapter 14
Robot Force Sensing Control
机器人的力觉控制

14.1 What is the force sensing control function? 什么是力觉控制功能?

After the robot is equipped with "force sensor", it performs the functions of pressure control, torque control, stiffness control, force detection, and force control data collection based on the received information of "force and torque" borne by each axial direction, which are called "force sensing control". The contents of "force sensing control" are as follows.

机器人在装备"力觉传感器"后,根据接收到的各轴向所承受的"作用力和力矩"等信息,执行压力控制、力矩控制、刚度控制、作用力检测、采集力觉控制数据等的功能,称之为"力觉控制"。"力觉控制"的内容如下。

1) Flexible control of the robot, according to the shape of the workpiece for action.

对机器人进行柔性控制,根据工件形状进行动作。

2) In any direction with a certain pressure to push the workpiece at the same time to move.

在任意方向上以一定压力推压工件的同时进行移动。

3) During the movement, the "stiffness" and "contact detection conditions" of the robot can be changed.

在动作过程中可以改变机器人的"刚度"及"接触检测条件"。

4) The ability to detect the contact state and perform interrupt insert processing using the contact state as a condition.

能够检测接触状态并以接触状态作为条件执行中断插入处理。

5) It can collect position information and force information when contacting objects.

能够采集接触对象物体时的位置信息及作用力信息。

6) Ability to collect force data synchronized with location data as log files.
能够采集与位置数据同步的作用力数据作为日志文件。

7) Ability to transfer log data files to FTP server.
能够将日志数据文件传送到 FTP 服务器。

14.2　Terms　术语

Stiffness control ——flexible control function of robot.
刚度控制——机器人柔性控制功能。

Force control —— the ability to push the workpiece with a specified force and control the movement of the robot.
作用力控制——以指定的力推压工件同时控制机器人移动的功能。

Interrupt insert run——the actual force and torque are detected to perform interrupt insert processing as a condition.
中断插入运行——检测实际作用力和力矩，以此作为条件，执行中断插入处理。

Data latching——the function of collecting and saving the force data and position data when contacting the workpiece object.
数据锁存——采集接触工件对象时的作用力数据、位置数据并保存的功能。

14.3　System composition　系统构成

（1）Composition of force sensing control system　力觉控制系统的构成（图 14-1）

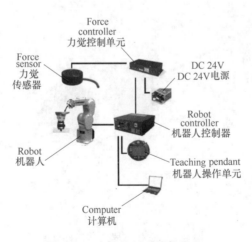

图 14-1　力觉控制系统的构成

（2）Product composition of force sensor controller　力觉控制器产品构成

Fig. 14-2 shows the product composition of force sensor controller.
图 14-2 所示是力觉控制器产品构成。

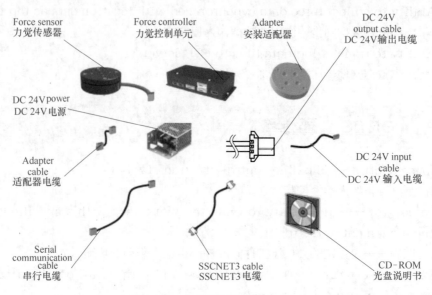

图 14-2　力觉控制器产品构成

(3) Force sensor installation components　力觉传感器安装部件

Fig. 14-3 is the decomposition diagram of force sensor installation components.

图 14-3 是力觉传感器安装部件分解图。

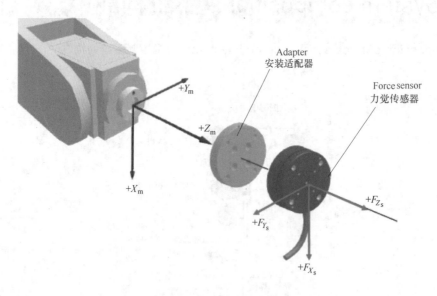

图 14-3　力觉传感器安装部件分解图

(4) Cable installation　电缆安装

Fig. 14-4 is the cable installation diagram.

图 14-4 是电缆安装图。

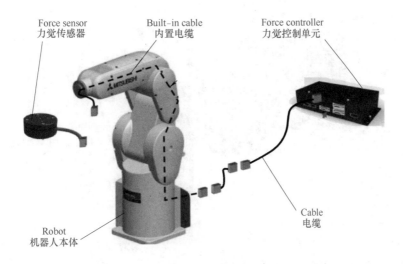

图 14-4　电缆从机器人本体内穿过

14.4　Force sense coordinate system
力觉坐标系

14.4.1　The definition of the force sense coordinate system
力觉坐标系的定义

The force sense coordinate system is used to calibrate the magnitude and direction of the force acting on the robot. The force sense coordinate system is the same as the coordinate system describing the position data. It's just that the "force sense coordinate system" demarcates "force and torque". There are several commonly used force sense coordinate systems.

力觉坐标系用于标定作用在机器人上的作用力大小和方向。力觉坐标系与描述位置数据的坐标系相同。只是"力觉坐标系"标定的内容为"作用力和力矩"。常使用的力觉坐标系有以下几种。

（1）Force sense mechanical interface coordinate system　力觉机械接口坐标系

"Force sense mechanical interface coordinate system" is based on the robot's IF coordinate system. The origin of "force sense mechanical interface coordinate system" coincides with the origin of the robot's IF coordinate system, but the direction of force and torque (+ direction) is shown in Fig. 14-5. F_{X_m}, F_{Y_m} and F_{Z_m} represent the force. Torque is represented by M_{X_m}, M_{Y_m} and M_{Z_m}.

"力觉机械接口坐标系"以机器人的 IF 坐标系为基准，"力觉机械接口坐标系"的原点与机器人的 IF 坐标系的原点重合，但作用力和力矩的方向（+向）如图 14-5 所示。F_{X_m}、F_{Y_m}、F_{Z_m} 表示作用力。M_{X_m}、M_{Y_m}、M_{Z_m} 表示力矩。

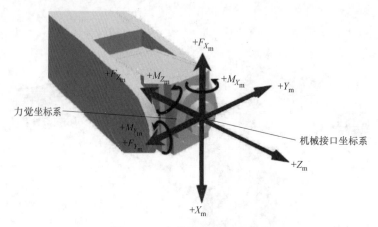

图 14-5　力觉机械接口坐标系

(2) Force sense tool coordinate system　力觉工具坐标系

The tool coordinate system of force sense is based on the "tool coordinate system" of the robot. The origin of the "force sense tool coordinate system" coincides with the

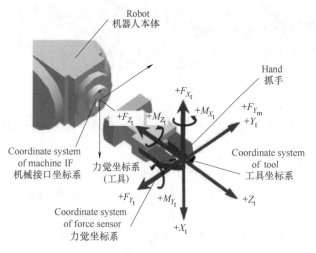

图 14-6　力觉工具坐标系

origin of the robot's "tool coordinate system", but the direction of force and torque (+direction) is shown in Fig. 14-6. F_{X_t}, F_{Y_t} and F_{Z_t} represent the force. Torque is expressed as M_{X_t}, M_{Y_t} and M_{Z_t}.

力觉工具坐标系以机器人的"工具(tool)坐标系"为基准。"力觉工具坐标系"的原点与机器人的"工具(tool)坐标系"的原点重合,但作用力和力矩的方向(+向)如图 14-6 所示。F_{X_t}、F_{Y_t}、F_{Z_t} 表示作用力。M_{X_t}、M_{Y_t}、M_{Z_t} 表示力矩。

(3) Force sense Cartesian coordinate system　力觉直角坐标系

"Force sense Cartesian coordinate system" is based on "robot Cartesian coordinate system". The origin of "force sense Cartesian coordinate system" coincides with the origin of "robot Cartesian coordinate system". The direction of force and torque (+direction) is shown in Fig. 14-7. F_X, F_Y, F_Z represent the force. Torque is expressed in M_X, M_Y and M_Z. The difference between "robot Cartesian coordinate system" and "robot mechanical interface coordinate system" lies in the position and direction of the X/Z axis.

"力觉直角坐标系"以"机器人的直角坐标系"为基准,"力觉直角坐标系"的原点与"机器人直角坐标系"的原点重合,作用力和力矩的方向(+向)如图 14-7 所示。F_X、F_Y、F_Z 表示作用力。M_X、M_Y、M_Z 表示力矩。"机器人直角坐标系"和"机器人机械接口坐标系"的区别在于 X/Z 轴的位置和方向不同。

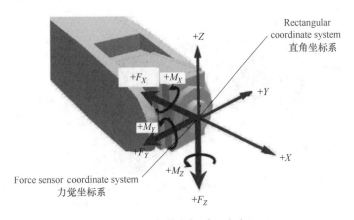

图 14-7　力觉坐标系（直交）

14.4.2　Force sensor coordinate system　力觉传感器坐标系

The force sensor coordinate system is based on the fixed position of the sensor. The origin of the force sensor coordinate system is at the center of the sensor's circle, and the direction of force and torque is shown in Fig. 14-8. F_{X_S}, F_{Y_S} and F_{Z_S} represent the acting force. The torque is represented by M_{X_S}, M_{Y_S} and M_{Z_S}. Fig. 14-9 shows the origin position of the force sensor coordinate system.

力觉传感器坐标系以传感器固定方位为基准。力觉传感器坐标系的原点位置在传感器的圆心位置，作用力和力矩的方向如图 14-8 所示。F_{X_S}、F_{Y_S}、F_{Z_S} 表示作用力。M_{X_S}、M_{Y_S}、M_{Z_S} 表示力矩。图 14-9 是力觉传感器坐标系的原点位置。

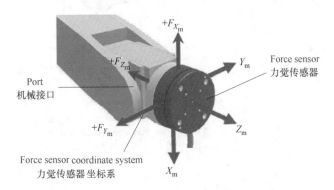

图 14-8　力觉传感器坐标系

As shown in Fig. 14-10, the position relationship between the mechanical coordinate system of force sense and the coordinate system of force sensor may not be consistent. If not, parameters need to be set for calibration.

如图 14-10 所示，力觉机械坐标系与力觉传感器坐标系的位置关系可能不一致，如果不一致时，需要设置参数进行标定。

The relationship between "force sensor coordinate system" and "force sense mechanical coordinate system" is shown in Fig. 14-11.

"力觉传感器坐标系"与"力觉机械坐标系"的关系如图 14-11 所示。

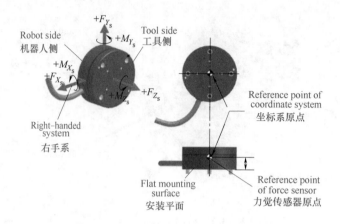

图 14-9　力觉传感器坐标系原点位置

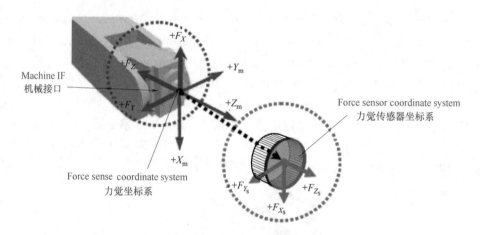

图 14-10　力觉机械坐标系与力觉传感器坐标系的位置关系

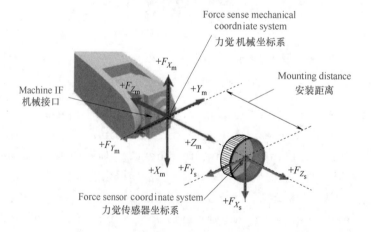

图 14-11　校正"力觉传感器坐标系"与"力觉机械坐标系"

14.5　Program　程序

14.5.1　Mode switching　模式切换

In the speed priority mode, the "speed command value" sets the "motion speed" when the robot is not in contact with the workpiece object. The automatic switch between "speed control" and "force control" is used to limit the speed at which the robot can approach the workpiece object to mitigate the impact. The switching conditions of "speed priority mode" and "force priority mode" are set by "mode switching judgment value" (Fig. 14-12).

在速度优先模式中，"速度指令值"设置机器人未与工件对象接触时的"运动速度"。使用"速度控制"和"作用力控制"自动切换，是为了限制机器人与工件对象的接近速度以缓和冲击。"速度优先模式"和"作用力优先模式"的切换条件由"模式切换判定值"设定（图 14-12）。

According to the value of force sensor, the priority mode is switched automatically. The critical value is "mode switching decision value".

根据力觉传感器数值，自动切换优先工作模式。临界值即"模式切换判定值"。

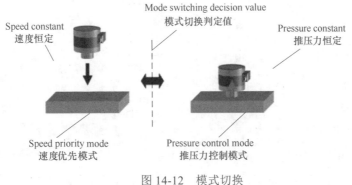

图 14-12　模式切换

Fig. 14-13 is a force control engineering case.

图 14-13 是作用力控制工程案例。

14.5.2　Sample program 1　样例程序 1

(1) Action description　工作要求

The robot moves along the Y axis, looking for objects. After the object is found, it keeps exerting a certain force in the Y axis direction of the object while moving in the X axis direction (Fig. 14-14).

机器人向 Y 轴方向动作，探寻对象物。发现对象物后，在对象物 Y 轴方向上保持施加一定的力，同时向 X 轴方向运动（图 14-14）。

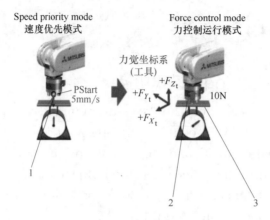

图 14-13　作用力控制工程案例

1—Move at a speed of 5mm/s before touching the workpiece——在接触工件前，以 5mm/s 的速度运动；
2—After contact with the workpiece, the force control mode is changed, press with 5N——接触工件后转变为"作用力控制模式"，以 5N 推压；
3—After contacting the workpiece and detecting force = 2.5N, switch to "thrust priority control mode" and push forward with 10N thrust——接触工件，检测到作用力 =2.5N 后，切换到"推力优先控制模式"，以 10N 推力推进

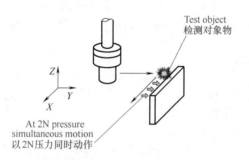

图 14-14　样例工程的动作过程

（2）Program　程序

```
'——设置控制模式 =0 的各技术参量。
P_FsStf0=（+0.00, +0.01, +0.00, +0.00, +0.00, +0.00）(0, 0)
                        '——刚度系数（设置 Y 轴 =0.01N/mm）。
P_FsDmp0=（+0.00, +0.00, +0.00, +0.00, +0.00, +0.00）(0, 0)
                        '——阻尼系数（不设定）。
P_FsMod0=（+0.00, +1.00, +0.00, +0.00, +0.00, +0.00）(0, 0)
                        '——力觉控制模式（设置 Y 轴 = 作用力
                          控制，因为需要保持 Y 向压力）。
M_FsCod0=1              '——设置采用力觉直角坐标系。
'——设置控制特性 =0 的各技术参量。
```

Chapter 14　Robot Force Sensing Control

```
P_FsGn0=(+0.00, +1.00, +0.00, +0.00, +0.00, +0.00)(0, 0)
                                    '——增益（设置 Y 轴 =1.0×10⁻³mm/N）。
P_FsFLm0=(+20.00, +0.50, +20.00, +5.00, +5.00, +5.00)(0, 0)
                                    '——设置作用力检测设定值（设置
                                       Y=0.5N）。
P_FsFCd0=(+0.00, +2.00, +0.00, +0.00, +0.00, +0.00)(0, 0)
                                    '——设置作用力指令（设置 Y=2.0N）。
'——主程序。
Def Act 1, M_FsLmtS=1 GoTo *XMOV, S  '——定义 Act 1：如果"作用力当前值"
                                       超过"检测设定值"的时候，跳转
                                       执行 *XMOV, S 中断程序。
FsLog On                             '——开始采集力觉日志数据。
Fsc On, 0, 0, 1                      '——力觉控制 ON. 控制模式组号 =0，控
                                       制特性组号 =0，零点清零。
P1=P_Curr                            '——设置 P1= 位置当前值。
P1.Y=P1.Y+200                        '——定义 P1.Y。
Spd 5                                '——设置速度。
Act 1=1                              '——Act 1 中断区间起点。
Mvs P1                               '——移动到 P1 点。在动作中如果检测到
                                       与对象物接触，则执行跳转处理。
Fsc Off                              '——力觉控制 OFF。
FsLog Off, 7                         '——结束力觉日志数据采集，创建日志
                                       文件 No.7。
Act 1=0                              '——Act 1 中断区间终点。
End                                  '——主程序结束。
'——中断程序。
*XMOV
P2=P_FsCurP                          '——设置 P2 为"力觉指令位置"。
P2.X=P2.X+100                        '——设置 P2.X 的位置。
FsGChg 5, 100, 2                     '——FsGChg 为切换控制特性指令，切
                                       换位置 =5%，切换时间 100ms，切
                                       换后控制特性组号 =2。
Mvs P2                               '——移动到 P2 点。在 Y 轴方向上保持施
                                       加 2.0N 的力，往 X 轴方向移动。
Fsc Off                              '——力觉控制 OFF。
FsLog Off, 7                         '——结束力觉日志数据的采集，创建日
                                       志文件 No.7。
FsOutLog 7                           '——通过 FTP 传送日志文件 No.7 到电
                                       脑上。
End                                  '——结束。
```

14.5.3　Sample program 2　样例程序2

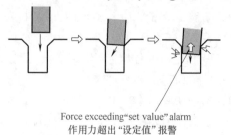

图 14-15　工作过程示意图

（1）Action description　工作要求

The robot works flexibly in XY direction and performs insertion in Z direction. If the force during insertion exceeds the set "check value", immediately alarm (Fig. 14-15).

机器人在 XY 方向处于弹性工作状态，在 Z 轴方向上执行插入动作。插入时如果作用力超过设置的"检测值"时，立即报警（图 14-15）。

（2）Program　程序

```
'——设置控制模式=0的各技术参量。
P_FsStf0=（+0.01, +0.01, +0.00, +0.00, +0.00, +0.00）(0, 0)
                '——刚度系数（设置X, Y=0.01N/mm）。
P_FsDmp0=（+0.00, +0.00, +0.00, +0.00, +0.00, +0.00）(0, 0)
                '——阻尼系数（不设置）。
P_FsMod0=（+2.00, +2.00, +0.00, +0.00, +0.00, +0.00）(0, 0)
                '——力觉控制模式（X, Y 轴=2：刚度控制）。
M_FsCod0=0      '——使用力觉工具坐标系。
'——设置控制特性=0的各技术参量。
P_FsGn0=（+80.00, +80.00, +0.00, +0.00, +0.00, +0.00）(0, 0)
```

'——增益（设置 X, Y 轴=80×10^{-3}mm/N）。

```
P_FsFLm0=（+20.00, +20.00, +10.00, +5.00, +5.00, +5.00）(0, 0)
                '——作用力检测设定值（设置Z 轴=10N）。
P_FsFCd0=（+0.00, +0.00, +0.00, +0.00, +0.00, +0.00）(0, 0)
                '——作用力指令（未设置）。
'——主程序。
Def Act 1, P_FsLmtX.Z=1 GoTo *ESCP, S
                '——定义 Act 1：如果Z 轴方向"实际作用力"
                   超过"检测设定值"时，跳转到中断程序
                   *ESCP, S。
Mov P1          '——移动到P1 点。
FsLog On        '——开始采集力觉日志数据。
Fsc On, 0, 0, 1 '——力觉控制 ON。控制模式组号=0, 控制特
                   性组号=0, 零点清零。
Spd 5           '——设置速度。
Act 1=1         '——Act 1 中断区间起点。
Mvs P2          '——移动到P2点。在Z轴方向上执行插入动作，
                   Z 轴上作用力如果超过"检测设定值"，
```

```
                            '——执行中断插入处理。
Fsc Off                     '——力觉控制 OFF。
FsLog Off, 1                '——结束力觉日志数据的采集,创建日志文件
                               No.1。
Act 1=0                     '——Act 1 中断区间终点。
End                         '——主程序结束。
'——中断插入程序。
*ESCP
Fsc Off                     '——力觉控制 OFF。
Spd 50                      '——设置速度。
Mvs P1                      '——退回到插入开始位置 P1。
FsLog Off, 1                '——结束力觉日志数据采集,创建日志文件
                               No.1。
FsOutLog 1                  '——通过 FTP 传送日志文件 No.1 到电脑上。
Error 9100                  '——报警 L9100 发生。
End                         '——程序结束。
```

14.5.4　Sample program 3　样例程序 3

（1）Action description　工作要求

The robot explores the holes in the XY plane. When the hole is found, calculate the XY coordinates for the center position. The action process is shown in Fig. 14-16.

机器人探索 XY 平面上的孔。发现孔的时候,算出该中心位置的 XY 坐标值。动作过程如图 14-16 所示。

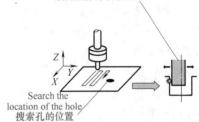

图 14-16　机器人探索 XY 平面上的孔

（2）Program　程序

```
'——设置控制模式 =0 的各技术参量。
P_FsStf0=（+0.00, +0.00, +1.00, +0.00, +0.00, +0.00）(0, 0)
                            '——刚度系数（设置 Z 轴 =1N/mm）。
P_FsDmp0=（+0.00, +0.00, +0.00, +0.00, +0.00, +0.00）(0, 0)
                            '——阻尼系数（未设置）。
P_FsMod0=（+0.00, +0.00, +2.00, +0.00, +0.00, +0.00）(0, 0)
                            '——力觉控制模式（设置 Z 轴为刚度控制）。
M_FsCod0=1                  '——设置采用力觉直角坐标系。
'——设置控制特性 =0 的各技术参量。
P_FsGn0=（+0.00, +0.00, +40.00, +0.00, +0.00, +0.00）(0, 0)
                            '——增益（Z 轴 =40×10$^{-3}$mm/N）。
```

```
P_FsFLm0=(+20.00, +20.00, +5.00, +5.00, +5.00, +5.00)(0, 0)
                          '——设置"作用力检测设定值"（设置Z轴=
                            5N）。
P_FsFCd0=(+0.00, +0.00, +0.00, +0.00, +0.00, +0.00)(0, 0)
                          '——作用力指令（未指定）。
'——主程序。
Def Act 1, P_FsLmtR.Z=1 GoTo *PCEN, S
                          '——定义Act 1：如果Z轴作用力大于"作用
                            力检测设定值"时，跳转执行中断程序
                            *PCEN, S。
P2=P1                     '——设置P2点。
P2.X=P2.X+100             '——设置P2点X值（X向探索行程100mm）。
Fsc On, 0, 0, 1           '——力觉控制ON。控制模式组号=0, 控制特
                            性组号=0, 零点清零。
Mvs P1                    '——移动到探索开始位置P1（探索平面在Z
                            向-10mm的位置，往Z方向推入）。
Spd 5                     '——设置速度。
Act 1=1                   '——Act 1中断区间起点。
For M1=1 To 10            '——Y方向上每隔5mm执行探索孔的位置。
Mvs P1                    '——移动到P1点。
Mvs P2                    '——移动到P2点。
P1.Y=P1.Y+5               '——设置P1.Y的数值。
P2.Y=P1.Y                 '——设置P2.Y的数值。
Mvs P2                    '——移动到P2点。
Mvs P1                    '——移动到P1点。
P1.Y=P1.Y+5               '——设置P1.Y的数值。
P2.Y=P1.Y                 '——设置P2.Y的数值。
Next M1
Act 1=0                   '——Act 1中断区间终点。
Fsc Off                   '——力觉控制OFF。
End                       '——主程序结束。
'——中断程序。
*PCEN
Dim PX(2), PY(2)          '——定义数组。
P0=P_Curr                 '——设置P0位置。以探索到孔，停止的位置
                            为基准。
PX(1)=P0                  '——设置PX(1)位置。
PX(2)=P0                  '——设置PX(2)位置。
PY(1)=P0                  '——设置PY(1)位置。
```

PY(2)=P0	'——设置PY(2)位置。
PX(1).X=P0.X+10	'——算出基准位置的XY方向上±10mm的位置。
PX(2).X=P0.X-10	'——设置PX(2)位置。
PY(1).Y=P0.Y+10	'——设置PY(1)位置。
PY(2).Y=P0.Y-10	'——设置PY(2)位置。
Fsc Off	'——力觉控制OFF。
P_FsFLm0=(+2.00,+2.00,+5.00,+5.00,+5.00,+5.00)(0,0)	'——X,Y轴方向的"作用力检测设定值"变更为2N。
Fsc On, 0, 0, 1	'——力觉控制ON。
MFLG=0	'——设置MFLG=0。
For M1=1 To 2	
Mvs PX(M1)WthIf P_FsLmtR.X=1, Skip	'——在向PX(M1)的移动过程中,如果X轴上的作用力超过"作用力检测设定值",就执行跳转。
IfM_SkipCq=1 Then	'——如果执行了跳转,则下一步。
PX(M1)=P_FsLmtP	'——设置PX(M1)保持为第一次"超过作用力检测设定值的位置"。
MFLG=MFLG+1	'——设置MFLG。
EndIf	
Mvs P0	'——移动到P0点。
Fsc Off	'——为了复位P_FsLmtP,使力觉控制从无效到有效。
Fsc On, 0, 0, 1	'——力觉控制ON。
Next M1	
For M1=1 To 2	
Mvs PY(M1)WthIf P_FsLmtR.Y=1, Skip	'——在向PY(M1)的移动过程中,如果Y轴上的作用力超过"作用力检测设定值",就执行跳转。
If M_SkipCq=1 Then	'——如果执行了跳转,则下一步。
PY(M1)=P_FsLmtP	'——跳转的时候,保持PY(M1)=超过作用力检测设定值的位置。
MFLG=MFLG+1	'——设置MFLG。
EndIf	
Mvs P0	'——移动到P0。
Fsc Off	'——力觉控制OFF。

```
Fsc On, 0, 0, 1                    '——为了复位 P_FsLmtP, 使力觉控制从无效
                                      到有效。
Next M1
If MFLG=4 Then                     '——找到 4 个点。
PTMP=(PX(1)+PX(2))/2               '——设置 PTMP。
P0.X=PTMP.X                        '——设定 X 轴方向的中心位置为 P0。
PTMP=(PY(1)+PY(2))/2               '——设置 PTMP。
P0.Y=PTMP.Y                        '——设定 Y 轴方向的中心位置为 P0。
Else                               '——没有找到 4 个点。
Error 9100                         '——报警 L9100。
EndIf                              '——结束。
End                                '——结束。
```

14.5.5　Sample program 4　样例程序 4

(1) Action description　工作要求

Moving at a specified speed in the Z axis, touching the workpiece object. After touching the workpiece, if the robot's Z-direction position and sensor data meet the specified conditions, keep pushing in the Z-direction and start to move in the Y axis (Fig. 14-17).

在 Z 轴方向上以指定的速度动作，接触工件对象物。接触工件后如果机器人的 Z 向位置和传感器数据满足所规定的条件，保持 Z 向的推入，Y 轴方向开始动作（图 14-17）。

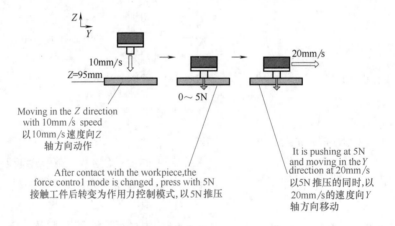

图 14-17　推压动作示意图

(2) Program　程序

```
'——设置控制模式 =0 的各技术参量。
M_FsCod0 = 0                              '——使用力觉工具坐标系。
P_FsMod0 = (0, 1, 1, 0, 0, 0)(0, 0)       '——设置 Y, Z 轴方向为"作用力控制"。
'——设置控制特性 =0 的各技术参量。
```

```
P_FsFCd0 = (0, 0, 5, 0, 0, 0)(0, 0)          '——设置Z轴方向作用力指令=5N。
P_FsSpd0 = (0, 0, 10, 0, 0, 0)(0, 0)         '——设置Z轴方向的速度=10mm/s。
P_FsSwF0 = (0, 0, 0.5, 0, 0, 0)(0, 0)        '——设置Z轴方向的模式切换判定值
                                               =0.5N。
P_FsGn0 = (0, 0, 10, 0, 0, 0)(0, 0)          '——设定Z轴方向的力觉控制增益
                                               =10×10⁻³mm/N。
'——为了使Y轴方向不动作,指定为0。
'——设置控制特性=-1的各技术参量。
P_FsFCd1 = (0, 3, 5, 0, 0, 0, 0, 0)(0, 0)    '——设置Y轴方向作用力指令=3N、Z
                                               轴方向=5N。
P_FsSpd1 = (0, 20, 0, 0, 0, 0, 0, 0)(0, 0)   '——设置Y轴方向的速度=20mm/s。
'——为了使Z轴方向继续执行作用力控制,设置0mm/s。
P_FsSwF1 = (0, 2, 0, 0, 0, 0, 0, 0)(0, 0)    '——设定Y轴方向模式切换判定值=2N
P_FsGn1 = (0, 10, 10, 0, 0, 0)(0, 0)         '——设定Y轴方向的力觉控制增益
                                               =10×10⁻³mm/N(往Y轴方向动作)。
'——主程序。
Def MoTrg 1,(P_Fbc.Z<100)And(P_FsCurD.Z>4.8)
                                             '——定义1#Mo 组合条件:如果Z向
                                               行程小于100同时Z向作用力大于
                                               4.8N。
SetMoTrg 1                                   '——设置1#Mo 组合条件有效。
Fsc On, 0, 0, 1                              '——力觉控制ON。控制模式组号=0、
                                               控制特性组号=0。
Wait M_MoTrg(1)=1                            '——如果组合条件1#Mo=ON,(表示
                                               在Z轴方向动作接触到对象物)。
Dly 3                                        '——(暂停)接触后,等待调整。
FsCTrg 1, 100, -1                            '——1#Mo =ON 后,切换控制特性,控
                                               制特性组号=-1。
'——Z轴方向上以5N的力推入的同时,在Y轴方向上开始动作。
```

14.6 Stiffness controlling 刚度控制

14.6.1 Stiffness controlling principle 刚度控制原理

"Stiffness control" is what makes a robot work as if it were loaded with a spring. "刚度控制"就是使机器人在受到外力后如同装上弹簧一样工作。

(1) Action description 工作要求

Stiffness control is required on F_{X_t} and F_{Y_t} axes. When an external force is applied to the X and Y planes, an elastic action is performed in the opposite direction of the external force. After the external force is removed, return to the original position (Fig. 14-18).

要求在 F_{X_t}、F_{Y_t} 轴上执行刚度控制。在 X, Y 平面上受到外力时，沿外力的反方向做弹性动作。外力解除后回到原来的位置（图 14-18）。

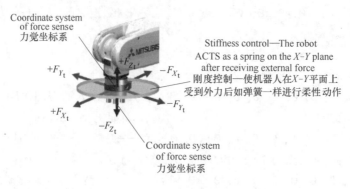

图 14-18 刚度控制

（2）Program　程序

```
'——设置控制模式组号 =0，使用变量设置下列参量。
P_FsStf0 = (+0.10, +0.10, +0.00, +0.00, +0.00, +0.00)(0, 0)
                              '——设置 X 轴、Y 轴的刚度系数 =
                                 0.1（N/mm）。重要！
P_FsDmp0 = (+0.00, +0.00, +0.00, +0.00, +0.00, +0.00)(0, 0)
                              '——根据现场状态设置阻尼系数。
P_FsMod0 = (+2.00, +2.00, +0.00, +0.00, +0.00, +0.00)(0, 0)
                              '——设置力觉控制模式（X, Y 轴：
                                 刚度控制）。
M_FsCod0 = 0                  '——采用力觉工具坐标系。
'——设置控制特性组号 =0，使用变量设置下列参量。
P_FsGn0 = (+20.00, +20.00, +0.00, +0.00, +0.00, +0.00)(0, 0)
                              '——设置 X、Y 轴增益 =20μm/N，
                                 需要根据现场状态进行调整。
P_FsFLm0 = (+0.00, +0.00, +0.00, +0.00, +0.00, +0.00)(0, 0)
                              '——设置作用力检测设定值（N）（本
                                 例中未设置）。
P_FsFCd0 = (+0.00, +0.00, +0.00, +0.00, +0.00, +0.00)(0, 0)
                              '——设置作用力指令（本例中未设置）。
P_FsSpd0 = (+0.00, +0.00, +0.00, +0.00, +0.00, +0.00)(0, 0)
                              '——设定速度控制模式的速度（本例
                                 中未设置）。
P_FsSwF0 = (+0.00, +0.00, +0.00, +0.00, +0.00, +0.00)(0, 0)
```

Chapter 14　Robot Force Sensing Control

```
                                          '——设定模式切换判定值（本例中未
                                             设置）。
' *** <刚度控制（X，Y 轴）程序> ***
Servo On                                  '——伺服 ON。
Mov PStart                                '——移动到起始点 PStart。
Dly 1                                     '——暂停。
Fsc On, 0, 0, 1                           '——"力觉控制 ON"。控制模式组号 =
                                             0，控制特性组号 =0，执行"清
                                             零"操作。
Hlt                                       '——暂停。
Fsc Off                                   '——力觉控制 OFF。
End                                       '——结束。
```

14.6.2　Change control characteristics 1　改变控制特性 1

As shown in Fig. 14-19:

如图 14-19 所示：

1）A robot can change the force as it moves;

机器人在运动过程中能够改变作用力；

2）start from P1 to P2 while pushing 5.0N in the Z direction;

在 Z 方向上以 5.0N 的力推压的同时，从 P1 向 P2 动作；

3）After the movement reaches 50%, gradually increase the thrust, and finally push the pressure with 12.0N force.

在移动行程到达 50% 位置开始逐渐加大推力，最终以 12N 的力推压。

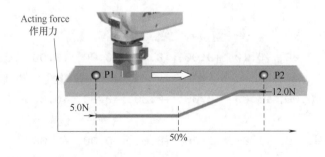

图 14-19　改变控制特性的工程案例

14.6.3　Change control characteristics 2　改变控制特性 2

（1）Action description　工作要求

As shown in Fig. 14-20, the robot moves along the guide rail.

如图 14-20 所示，机器人沿着导轨动作。

1）While pushing in with a force of 5N in the −X direction, it starts at a speed of 10mm/s in the +Y direction.

在 $-X$ 方向以 5N 的力推入的同时，在 $+Y$ 方向上以 10mm/s 的速度开始动作。

2）When position A detects contact with the workpiece surface in the $+Y$ direction, switch to push in with a force of 5N in the $+Y$ direction and move at a speed of 10mm/s in the $+X$ direction.

在位置 A 检测到与 $+Y$ 方向的工件面接触时，切换为在 $+Y$ 方向以 5N 的力推入，同时在 $+X$ 方向上以 10 mm /s 的速度动作。

3）Stop at position B when it detects contact with the workpiece surface in the $+X$ direction.

在位置 B 检测到与 $+X$ 方向的工件面接触后停止动作。

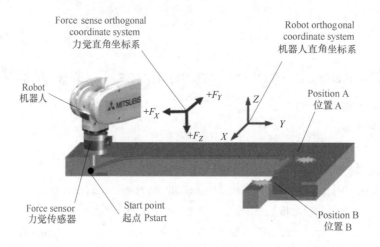

图 14-20　改变控制特性的工程案例 2

（2）Program　程序

```
'——选择控制模式组号 =0，设置技术参量。
P_FsStf0 = (+0.00, +0.10, +0.00, +0.00, +0.00, +0.00)(0, 0)
                                '——设置 Y 轴刚度系数 =0.1N/mm。
P_FsDmp0 = (+0.00, +0.00, +0.00, +0.00, +0.00, +0.00)(0, 0)
                                '——设置阻尼系数 =0。
P_FsMod0 = (+1.00, +1.00, +0.00, +0.00, +0.00, +0.00)(0, 0)
                                '——设置作用力控制模式（X、Y 轴：
                                   作用力控制）。
M_FsCod0 = 1                    '——采用"力觉直角坐标系"。
'——设置"控制特性"组号 =0 的技术参量。
P_FsGn0 = (+2.00, +2.00, +0.00, +0.00, +0.00, +0.00)(0, 0)
                                '——设置 X、Y 轴增益 =2.0μm/N。
P_FsFLm0 = (+0.00, +0.00, +0.00, +0.00, +0.00, +0.00)(0, 0)
```

Chapter 14　Robot Force Sensing Control

```
                                                '——设置作用力检测设定值（N）（未
                                                   设置）。
P_FsFCd0 =（-5.00, +5.00, +0.00, +0.00, +0.00, +0.00）(0, 0)
                                                '——设置作用力指令值X轴=-5.0N，
                                                   Y轴=5.0N。
P_FsSpd0 =（+0.00, +10.00, +0.00, +0.00, +0.00, +0.00）(0, 0)
                                                '——设置速度。
P_FsSwF0 =（+0.00, +3.00, +0.00, +0.00, +0.00, +0.00）(0, 0)
                                                '——设置模式切换判定值。
'——设置控制特性组号=-1的各技术参量。
P_FsGn1 =（+2.00, +2.00, +0.00, +0.00, +0.00, +0.00）(0, 0)
                                                '——设置增益。
P_FsFLm1 =（+0.00, +0.00, +0.00, +0.00, +0.00, +0.00）(0, 0)
                                                '——设置作用力检测设定值。
P_FsFCd1 =（+5.00, +5.00, +0.00, +0.00, +0.00, +0.00）(0, 0)
                                                '——设置作用力指令值X轴=5.0N，
                                                   Y轴=5.0N。
P_FsSpd1 =（+10.00, +0.00, +0.00, +0.00, +0.00, +0.00）(0, 0)
                                                '——设定速度控制模式的速度。
P_FsSwF1 =（+3.00, +0.00, +0.00, +0.00, +0.00, +0.00）(0, 0)
                                                '——设置模式切换判定值。
'*** <作用力控制-改变作用力方向的程序> ***
Def MoTrg 1, P_FsCurD.Y>4.5           '——定义#1Mo组合条件：力觉传
                                           感器数据 Fy > 4.5N。
Mvs PStart                             '——向初始位置点PStart移动。
Dly 1                                  '——暂停，等待直到完全静止。
Fsc On, 0, 0, 1                        '——力觉控制ON。
FsCTrg 1, 100, -1, 30, 0 , 1           '——根据#1Mo组合条件，指定控制
                                           特性切换；切换时间为100ms；
                                           切换后的"控制特性组号=-1"；
                                           "超时时间"设置=30s；超时不
                                           报警。
*LBL1: If P_FsCurD.X<4.5 Then GoTo *LBL1
                                        '——如果FX<4.5N，就一直等待；
                                           如果FX>4.5N，就进入程序下
                                           一行。
Fsc Off                                 '——力觉控制OFF。
End                                     '——结束。
```

14.7　Assembly　装配

14.7.1　Case 1　案例 1

（1）Action description　工作内容

As shown in Fig. 14-21, when the robot detects an abnormal impact force such as a collision, it stops immediately.

如图 14-21 所示，机器人在检测到发生碰撞等异常的冲击力时立即紧急停止。

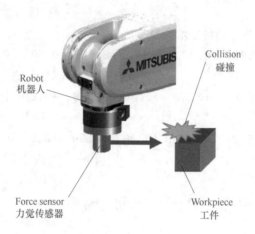

图 14-21　机器人在检测到异常冲击力时紧急停止

（2）Program　程序

```
'——设置控制模式组号 =0 时的各技术参量。
P_FsStf0 =（+0.00, +0.00, +0.00, +0.00, +0.00, +0.00）(0, 0)
                                    '——刚度系数 =0N/mm。
P_FsDmp0 =（+0.00, +0.00, +0.00, +0.00, +0.00, +0.00）(0, 0)
                                    '——阻尼系数 =0。
P_FsMod0 =（+0.00, +0.00, +0.00, +0.00, +0.00, +0.00）(0, 0)
                                    '——力觉控制模式（全轴作位置控制）。
M_FsCod0 = 0                        '——采用力觉工具坐标系。
'——设置控制特性组号 =0 时的各技术参量。
P_FsGn0 =（+0.00, +0.00, +0.00, +0.00, +0.00, +0.00）(0, 0)
                                    '——增益 =0μm/N。
P_FsFLm0 =（+50.00, +50.00, +50.00, +0.50, +0.50, +0.50）(0, 0)
                                    '——设置各轴"作用力检测设定值"。
P_FsFCd0 =（+0.00, +0.00, +0.00, +0.00, +0.00, +0.00）(0, 0)
                                    '——作用力指令值（未设置）。
P_FsSpd0 =（+0.00, +0.00, +0.00, +0.00, +0.00, +0.00）(0, 0)
                                    '——设定速度（未设置）。
```

```
P_FsSwF0 = (+0.00, +0.00, +0.00, +0.00, +0.00, +0.00)(0, 0)
                                         '——设定模式切换判定值。
'*** <主程序> ***
Def Act 1, M_FsLmtS=1 GoTo *XERR, F      '——如果状态变量 M_FsLmtS 的值 =1
                                         （超过"作用力检测设定值"），则
                                         跳转执行中断程序"XERR"。
Spd 10                                   '——设置速度。
Fsc On, 0, 0, 1                          '——力觉控制 ON。
Act 1=1                                  '——设置"中断区间有效起点"。
Mvs P1                                   '——移动到 P1 点。
Fsc Off                                  '——力觉控制 OFF。
End
*XERR                                    '——子程序号。
Act 1=0                                  '——退出 Act 1 中断区间。
Error 9100                               '——报警发生。
End                                      '——结束。
```

14.7.2 Case 2 案例 2

(1) Action description 工作内容

As shown in Fig. 14-22, the push force and position of the robot are detected at the same time to determine whether the assembly work is completed.

如图 14-22 所示，同时检测机器人推入力和位置、判定装配工作是否完成。

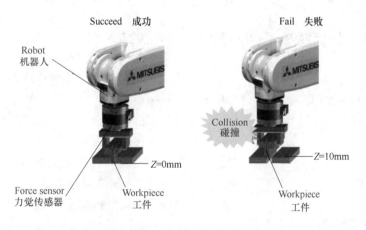

图 14-22 位置及作用力检测

1) The assembly operation is judged to be successful when the force in the direction of F_{Z_t} exceeds +18N and the Z coordinate is less than 5mm.

在 Z 坐标 5mm 以下、并且检测到在 F_{Z_t} 方向作用力超过 +18N 时判断为装配作业成功。

2) If the above conditions are not met within 5 seconds after the installation

begins, the installation is judged to have failed and an alarm is output.
安装开始后5s内不满足上述条件时，判断为安装失败并输出报警。

（2）Program　程序

```
'——设置控制模式组号 =0 时的各技术参数。
P_FsStf0=（+0.00, +0.00, +0.00, +0.00, +0.00, +0.00）(0, 0)
                              '——设置刚度系数（N/mm）。
P_FsDmp0=（+0.00, +0.00, +0.00, +0.00, +0.00, +0.00）(0, 0)
                              '——设置阻尼系数。
P_FsMod0=（+2.00, +2.00, +1.00, +0.00, +0.00, +0.00）(0, 0)
                              '——设置力觉控制模式（X、Y：刚度控制；
                                Z：作用力控制）。
M_FsCod0 = 0                  '——采用力觉工具坐标系。
'——设置控制特性组号 =0 时的各技术参数。
P_FsGn0 =（+2.00, +2.00, +2.00, +0.00, +0.00, +0.00）(0, 0)
                              '——设置增益（μm/N）。
P_FsFLm0=（+0.00, +0.00, +0.00, +0.0, +0.0, +0.0）(0, 0)
                              '——设置作用力检测设定值（N）。
P_FsFCd0=（+0.00, +0.00, +20.00, +0.00, +0.00, +0.00）(0, 0)
                              '——设置作用力指令值（Z 轴 =20N）。
P_FsSpd0=（+0.00, +0.00, +0.00, +0.00, +0.00, +0.00）(0, 0)
                              '——设定速度。
P_FsSwF0=（+0.00, +0.00, +0.00, +0.00, +0.00, +0.00）(0, 0)
                              '——设置模式切换判定值。
'*** <主程序> ***
Def MoTrg 1,（(P_Fbc.Z <= 5) AND (P_FsCurD.Z > 18)）
                              '——定义 #1 Mo 组合条件（Z 轴位置
                                <5mm 以及 Z 向作用力 >18N）。
Def Act 1, M_MoTrg (1)=1 GoTo *XOK, F
                              '——设定 #1 Mo 组合条件 =1，就跳转执
                                行 "XOK, F" 中断程序。
Mvs PStart                    '——移动到 PStart 位置。
SetMoTrg 1                    '——设置 #1Mo 组合条件有效。
Fsc On, 0, 0, 1               '—— "力觉控制 ON"。
Act 1=1                       '——进入中断有效区间。
M_Timer (1) =0                '——定时器清零。
*LBL1: If M_Timer (1) < 5000 Then GoTo *LBL1
                              '——判断：如果工作时间大于 5s，则执行
                                下一行。
Fsc Off                       '——力觉控制 OFF。
Error 9100                    '——报警。
```

```
End                         '——结束。
*XOK                        '——"XOK"中断程序（如果装配完成，
                                就回程）。
Act 1=0                     '——中断区间无效。
SetMoTrg 0                  '——设置 Mo 组合条件无效。
Fsc Off                     '——力觉控制 OFF。
HOpen 1                     '——张开抓手。
P2=P_Fbc                    '——获取反馈位置，设置 P2 点。
P2.Z=P2.Z+100               '——在当前的位置以 Z 方向 +100mm 的位
                                置作为目标位置。
Mvs P2                      '——移动到 P2 点。
End                         '——结束。
```

14.7.3　Case 3: data latching and reading　案例 3：数据锁存及读取

As shown in Fig. 14-23:

如图 14-23 所示：

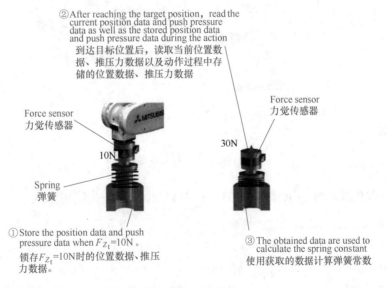

② After reaching the target position, read the current position data and push pressure data as well as the stored position data and push pressure data during the action

到达目标位置后，读取当前位置数据、推压力数据以及动作过程中存储的位置数据、推压力数据

Force sensor 力觉传感器

Spring 弹簧

① Store the position data and push pressure data when F_{Z_t}=10N。

锁存 F_{Z_t}=10N 时的位置数据、推压力数据。

③ The obtained data are used to calculate the spring constant

使用获取的数据计算弹簧常数

图 14-23　锁存及读取数据

1）According to the position data and the force data in the latching push action, calculate the spring constant of the spring component.

根据锁存推入动作中的位置数据和作用力数据，计算弹簧部件的弹簧常数。

2）It is required to save the position data and force data of each axis at the time point when F_{Z_t} exceeds 10N（latching function）.

要求在 F_{Z_t} 超出 10N 的时间点，保存各轴的位置数据和作用力数据（锁存功能）。

3) It can make the robot move to the final push position and calculate the spring constant according to the current latching position data and force data.

能使机器人运动到最终的推入位置,根据锁存的当前位置数据和作用力数据计算弹簧常数。

14.7.4　Data transfer case　数据传送案例

Transfer the collected log data files to PC(FTP server)。

将采集的日志数据文件传送到计算机(FTP 服务器)。

1) Collect the force sense log data in force sense control.

采集力觉控制中的力觉日志数据。

2) It will save the force sense log data, saved log files to the PC.

将保存力觉日志数据、保存的日志文件传送到计算机。

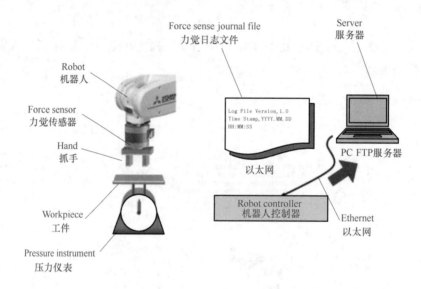

图 14-24　采集及传送数据

14.7.5　Position alignment push in　位置对准推入

(1) Action description　工作内容

Explore the phase of the gear and insert the metal shaft(Fig. 14-25)。

探索齿轮的相位,插入金属轴(图 14-25)。

1) By stiffness control, the robot is in a flexible working state, rotating in the direction of C axis while being pushed gently in the Z direction.

通过刚度控制使机器人处于柔性工作状态,在 Z 向轻轻推入的同时,在 C 轴方向旋转。

2) After the phase alignment of the gear and the metal shaft, the torque(M_z) about the Z axis will increase. Execute program jump when M_z increase is detected, as shown in Fig. 14-25.

齿轮和金属轴的相位对准后,绕 Z 轴的力矩(M_z)将增加。当检测到 M_z 增加时执行程序跳转,如图 14-25。

Chapter 14　Robot Force Sensing Control

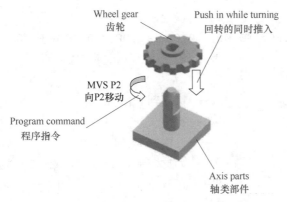

图 14-25　相位对准插入工程

（2）Program　程序

```
'——设置控制模式 =0 的各技术参量。
P_FsStf0=（+0.00，+0.00，+3.00，+0.00，+0.00，+0.50）(0，0)
                    '——刚度系数 N/mm（设置 Z 轴、C 轴刚度系数）。
P_FsDmp0=（+0.00，+0.00，+0.00，+0.00，+0.00，+0.00）(0，0)
                    '——阻尼系数。
P_FsMod0=（+2.00，+2.00，+2.00，+0.00，+0.00，+2.00）(0，0)
                    '——设置力觉控制模式（X、Y、Z、C：刚度控制）。
M_FsCod0 = 0        '——采用力觉工具坐标系。
'——设置控制特性 =0 的各技术参量。
P_FsGn0 =（+2.50，+2.50，+2.50，+0.00，+0.00，+2.50）(0，0)
                    '——设置增益（μm/N）。
P_FsFLm0=（+0.00，+0.00，+0.00，+0.00，+0.00，+0.05）(0，0)
                    '——作用力检测设定值（N）（$M_z$=0.05N·m）。
P_FsFCd0=（+0.00，+0.00，+0.00，+0.00，+0.00，+0.00）(0，0)
                    '——设置作用力指令。
P_FsSpd0=（+0.00，+0.00，+0.00，+0.00，+0.00，+0.00）(0，0)
                    '——设置速度。
P_FsSwF0=（+0.00，+0.00，+0.00，+0.00，+0.00，+0.00）(0，0)
                    '——设置模式切换判定值。
'——设置控制特性 =-1 的各技术参量。
P_FsGn1 =（+2.50，+2.50，+0.00，+0.00，+0.00，+3.00）(0，0)
                    '——设置增益（μm/N）（Z 轴 =0.0μm/N）。
P_FsFLm1=（+0.00，+0.00，+0.00，+0.00，+0.00，+0.00）(0，0)
                    '——作用力检测设定值（N）。
P_FsFCd1=（+0.00，+0.00，+0.00，+0.00，+0.00，+0.00）(0，0)
                    '——设置作用力指令。
P_FsSpd1=（+0.00，+0.00，+0.00，+0.00，+0.00，+0.00）(0，0)
```

```
                                        '——设定速度。
P_FsSwF1=(+0.00, +0.00, +0.00, +0.00, +0.00, +0.00)(0, 0)
                                        '——设置模式切换判定值。
'*** <装配作业程序> ***
Mvs PStart                              '——移动到 PStart 点。
Dly 1                                   '——暂停。
Ovrd 5                                  '——设置速度倍率。
Fsc On, 0, 0, 1                         '——力觉控制 ON。X、Y、Z、C 轴为刚度控制模式。
Mvs P1                                  '——运动到 P1 点。P1 点位于装配开始位置以下
                                           1mm。
Mvs P2 Wthif P_FsLmtR.C>0, Skip         '——向 P2 点移动（带动齿轮旋转）。如果移动过程
                                           中 C 轴力矩 $M_z \geqslant 0.05\text{N}\cdot\text{m}$，则 Skip（跳到
                                           下一行）。
If M_SkipCq = 0 Then *LERR              '——如果未执行 Skip 跳转，则跳转到 *LERR 标
                                           签处。
FsGChg 0, 10, -1                        '——控制特性"组号"变更为"-1"（位置控制）。
Mvs , 10                                '——向 +Z 上运动 10mm。
HOpen 1                                 '——张开抓手。
Fsc Off                                 '——力觉控制 OFF。
Mvs PStart                              '——运动到 PStart。
End                                     '——结束。
*LERR                                   '——报警处理。
Error 9100                              '——报警 9100。
End                                     '——结束。
```

References
参考文献

[1] 黄风. 工业机器人与自控系统的集成应用[M]. 北京：化学工业出版社，2017.
[2] 黄风. 工业机器人编程指令详解[M]. 北京：化学工业出版社，2017.
[3] 陈先锋. 伺服控制技术自学手册[M]. 北京：人民邮电出版社，2010.
[4] 杨叔子，杨克冲，吴波，等. 机械工程控制基础[M]. 7版. 武汉：华中科技大学出版社，2017.